Seeding the Universe with Life

~ Securing Our Cosmological Future ~

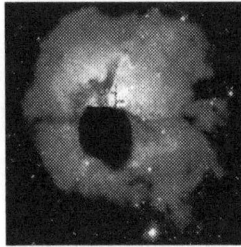

- Galactic Ecology, Astroethics
and Directed Panspermia -

Michael Noah Mautner   Ph.D.

Cover Photo: The first extraterrestrial plant, a small potato plant grown on meteorite extracts. The plant, positioned on its parent meteorite, is pictured sailing toward new suns. Astroecology studies showed that similar asteroid resources can support immense biological populations in the Solar System, and throughout the galaxy.

Legacy **LB** **Books**

Legacy Books Ltd
P.O. Box 7465 Christchurch, New Zealand
Published simultaneously on the Internet www.Legacy-Books.com and www.ebookmall.com

Contacts: The Society for Life in Space (SOLIS)

www.panspermia-society.com E-mail solis@solis1.com

Contents

Plant culture grown on meteorite soil

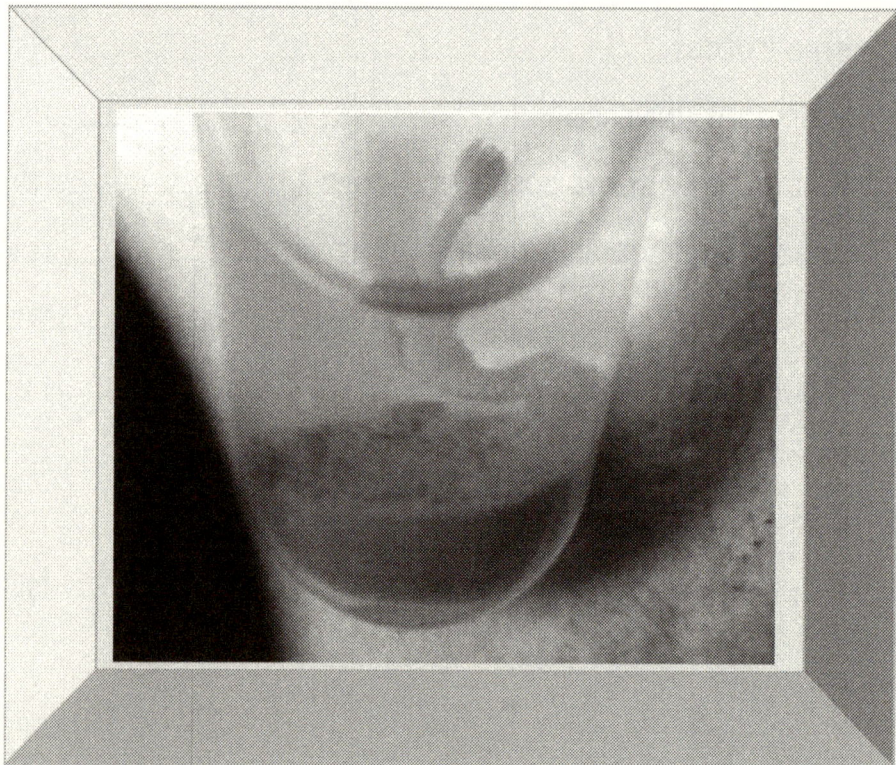

*L*ife is unique in the universe, as the laws of nature precisely allow living patterns to exist. The universe therefore has come to a unique point in Life.

*L*ife creates complex structures, and central to Life is propagating its patterns. Where there is life, there is purpose.

*B*eing part of life defines the human purpose to forever protect and propagate life.

*T*his purpose defines a life-centered ethics. That which promotes life is good, and that which harms life is evil. A life-centered panbiotic ethics seeks to expand life in the universe.

*W*e are now able to advance life on a cosmic scale. We can seed new solar systems with organisms that bear gene/protein life. Some will advance into intelligent beings who will then propagate life further in the galaxy.

*T*he main resources in space will be asteroids and comets. We examined these materials in meteorites and found them to be fertile.

*U*sing astro-ecology and cosmology, we can calculate the amounts of future life that the galaxy can support. The potential of future life is immense.

*I*f we fulfil our purpose, life will fill the universe. In our descendants throughout the universe, our human existence will find a cosmic purpose.

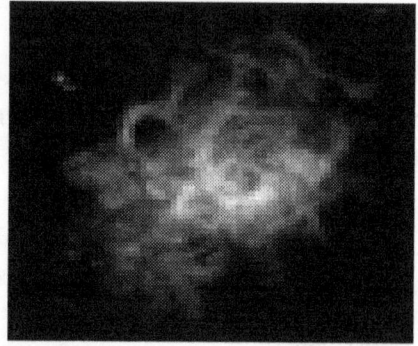

Star - forming interstellar clouds: the targets

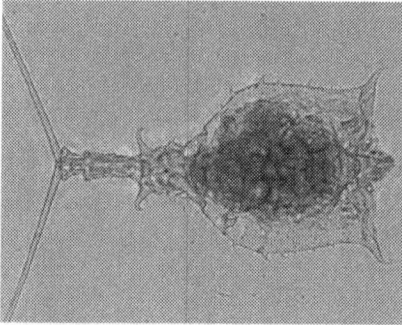

Multicellular rotifers - towards higher evolution

Comets - vehicles of storage and delivery

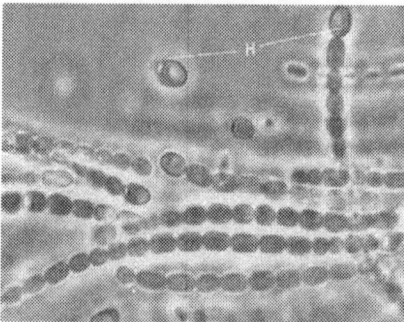

Blue-green algae - the first colonizers

Astrobiota

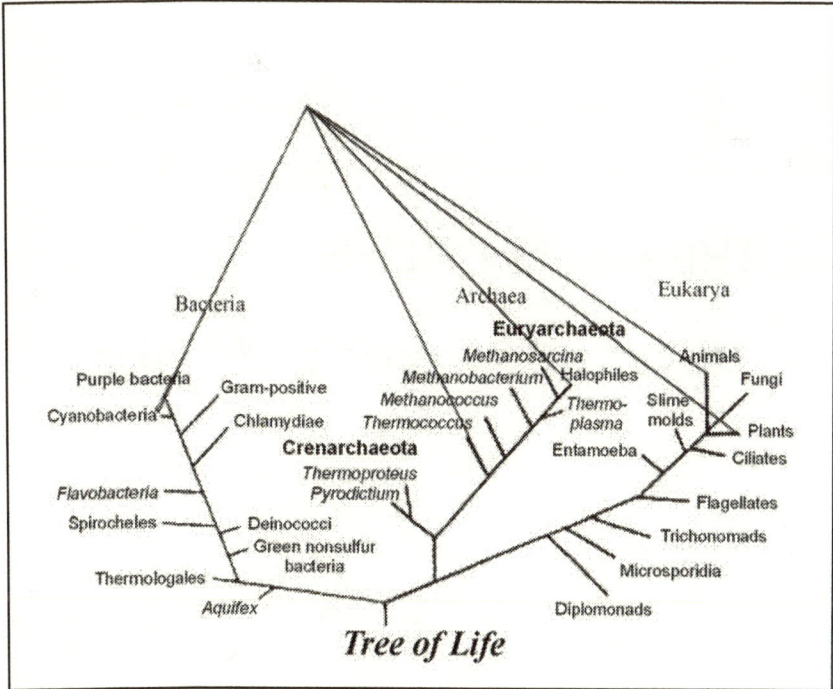

Tree of Life

The phylogenetic tree of life. Species that diverged from a common ancestor will converge again in space populations. Life in space will combine features from all the living kingdoms, and develop new features adapted to new environments. This new biota that pervades space may be called astrobiota, or panbiota.

1

THE FAMILY OF LIFE:

Its Nature, Purpose and Ethics

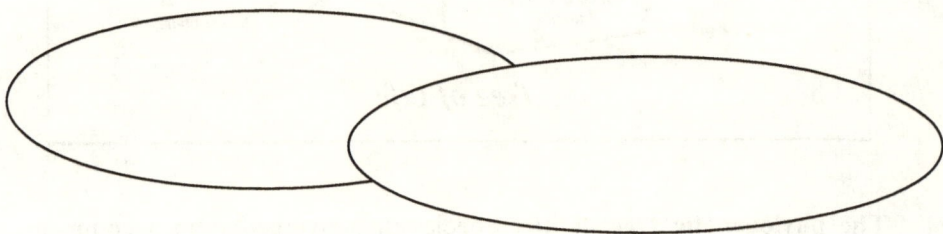

Chapter 1.1 The Family of Life: a Common Purpose and a Shared Future

*W*e share with all life the common structures of genes and proteins. They indicate the common origins of life, and imply a shared future. As long as we stay human, we shall carry the essence of life in our every cell.

What is life? What structures of Nature should we accept as fellow life? These are questions for human judgement, and science can help to formulate the answers.

Life is a process wrought in self-propagating molecular structures. Indeed, any observer would note that living beings focus on self-propagation. If not consciously, then by observational equivalence, the acts of life pursue this purpose. Where there is life, there is purpose; and a universe that contains life, contains purpose.

We are part of Life and share its purpose. The human purpose is therefore to forever safeguard and propagate life, and to assure that life will fill the universe.

Since its origins life has been subject to a simple logic. Those who survive and multiply inherit the future; those who do not, vanish. This self-evident logic has directed life to develop increasingly fit, advanced life-forms.

The logic of life assures that those who seek propagation most, will succeed best. Life will always be subject to this logic. Therefore, we must assure that our descendants will always desire self-propagation. This will be assured if control remains in organic brains with a vested interest to continue gene/protein life.

Self-propagation will then drive our descendants to expand in the universe. Space is replete with resources, and can harbor immense amounts of life for many eons.

The diversity of life is practically limitless. Expansion into space will require life to assume ever newer patterns, and space can sustain ever more advanced life-forms.

We can start to perceive our cosmological future. We can reach for this future by populating our Solar System and by seeding other solar systems. Our family of gene/protein life will realize its full potential when our descendants seek a cosmological future.

Chapter 1.2 Life and Purpose are Defined by Human Judgement

"Life" is defined by human judgement. More precisely, when we seek to define Life we consider the question "what is common to those whom we consider as fellow living beings". This is a question of human perceptions.

Many sensible definitions of Life have been suggested. However, Nature creates objects and patterns but does not call them names. Only humans use definitions. Therefore, "what is Life" cannot be answered by Nature. It must be answered by human beings using human faculties: our instincts, emotions and rational arguments. Whomever we accept as living beings – they *are* then by definition, part of Life.

First, we can use our innate feelings. All living beings have a special relationship with each other. It is obvious to every child and to all living things at some level, who is a fellow living being. Second, we can use a deeper understanding derived from science. All whom our innate feelings consider as living beings reproduce through genes made of DNA sequences that code proteins which directly or indirectly reproduce the DNA code. This cycle is at the heart of life as we commonly know it.

Importantly, by pursuing this core cycle, *life is a process.* Collections of inactive DNA, isolated proteins or inert organisms are not alive. The living process involves action and change. This has profound consequences. For example, life cannot exist in an environment of equilibrium where nothing can happen. By its nature, the living process involves the flow of matter and energy to maintain complex structures that would otherwise dissipate, due to statistics that disfavors organized structure. Living structures must be always maintained actively.

The central activities of Life favor a selected outcome, self-propagation. The actions of living beings will appear to any observer as if they purposefully seek self-propagation. Of course, only conscious humans seek objectives with foresight. Living beings act in all observable aspects as if they seek self-propagation purposefully. In turn, modern science holds that if two items are equivalent in all observable properties, then they are in fact identical. By the criteria of observational science, the living process does in effect seek a purpose.

Life is the self-propagation of organic matter through gene/protein cycles. The effective purpose of Life is to continue to live. Where there is Life, there is a purpose.

4

Chapter 1.3 The Unity of Life

Common Structures, Common Origins, and a Linked History

All cells are made of proteins with complex structures that are suited to specific functions. Particularly important to self-propagation are the enzymes that translate DNA codes into protein sequences, and the t-RNA molecule that associates each DNA triplet with an amino acid that is added to the protein when it is constructed. The proteins, in turn, help to reproduce the DNA code. Every species from microbes to humans shares the same complex mechanism of reproduction. They all contain thousands of complex enzymes tailored specifically for their functions. Only living beings in all of Nature possess such complex and specific structures.

Molecular evolution suggests that the three main branches of life – eubacteria, nucleated eukaryotes and archae all evolved from a common ancestor or group of ancestors. At the simplest early stage, protocells were made of loosely bound components who easily exchanged genes. In fact, all living beings were then one genetically coupled organism. The phylogenetic tree, based on sequences of bases in DNA and sequences of amino acids in proteins, show that all branches of life are related.

Today, genes still flow freely within each species and are sometimes transferred amongst different species. Genetic advances can be distributed in the family of life.

The evolution of all species influenced each other through cooperation, symbiosis, competition, ecological webs and genetic exchanges. Their history is recorded in fossils and in the similarities and differences of the anatomy of organisms. These visible signs are confirmed by the relations of DNA and protein sequences that lead to the phylogenetic tree of Life.

Common origins, shared gene/protein structures, the interchange of genes, shared ecology and resources, judgement by survival, and a shared future – all unite our family of organic Life.

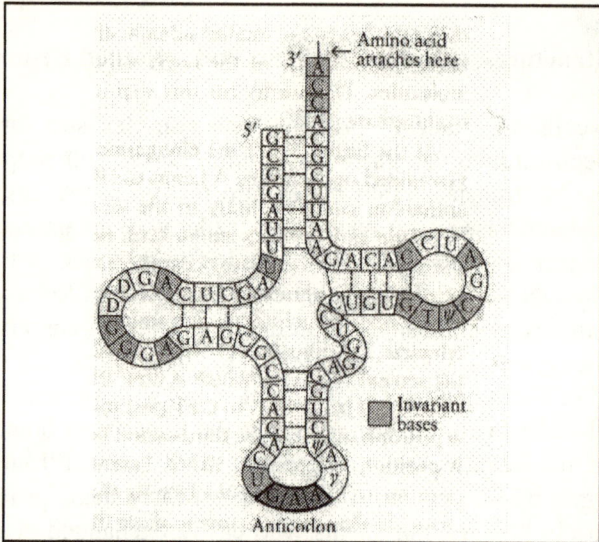

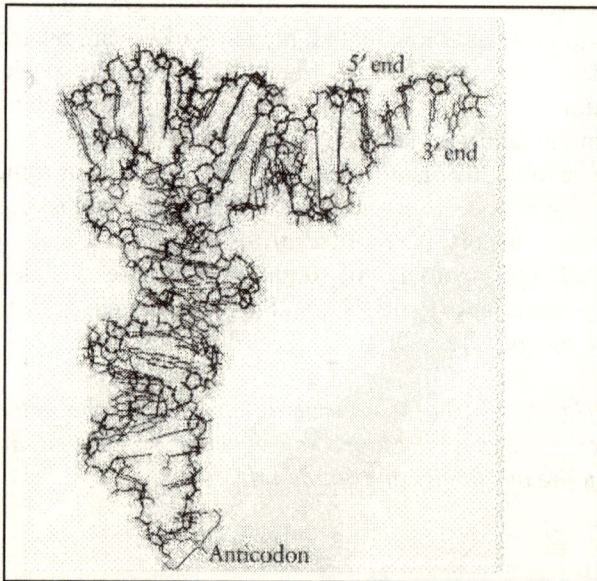

Two representations of the tRNA molecule that translate the genetic code into protein sequences. This molecule is central to the gene/protein cycle. It is one of the many complex molecules that are shared by all living cells. These shared features indicate a common ancestor.

A Shared Future

As Life needs to adapt to diverse environments in space, traits from very different species will become useful in new combinations. For example, self-contained humans living in space will need photosynthesis to create their own food internally. This can be achieved by implanting algae into human skin or by creating new photosynthetic human organs. Other plant genes can be also used by humans to provide waxy skin that prevents the loss of gas and water in space. Traits to withstand heat, acids and bases, dryness and radiation can be obtained from extremophile microorganisms and incorporated into higher plants, animals and humans to adapt to extreme environments. In this manner, sharing genes amongst species will unite future life.

Microorganisms pioneered life on Earth and similarly, they can pioneer life on new planets and establish ecosystems suitable for humans. Complex new ecosystems will need diverse species to exist together to maintain livable conditions and to recycle resources. This will unite the future of the various species.

New forms of life may evolve naturally and some will also be engineered by our descendants. These new species will coexist, support each other, build ecosystems, and co-evolve together.

To survive and reproduce, future species will have to actively maintain organic life. They will always be subject to the Logic of Life, the tests of survival. This too will unite the future of all living beings.

Future life will combine the genetic heritage of many diverse species. New species will emerge, coexist, and support each other. However it may evolve, all forms of life will be subject to the tests of survival. The patterns of Life may forever change, but the logic of Life is forever permanent.

Whom Should We Accept as Fellow Life?

Having identified the common features of living beings, it is inviting to find a deeper definition of life. We may call life all that self-propagates and contains structure and information, or only cellular, organic gene/protein organisms like ourselves. We shall propagate with whomever we recognize as fellow life, and therefore these gene-protein patterns will multiply in nature. Our definition of life can therefore have profound effects on the future.

All cellular gene-protein organisms belong to our family of life. However, we are already at the brink of altering genes and proteins, extending them with synthetic components that are optimized for new functions. This modified biology may adapt better better adapted to new space environments. They may be cellular gene/protein life made of organic molecules, cells and organs, but having a different chemistry than ours. We may also encounter such organisms with variant gene/protein chemistry that evolved naturally. Should we accept these new life forms as fellow life?

A general definition of life may also include non-biological robots. We will construct robots for good reasons, as they can serve us. They can explore space because robots can greatly exceed us in endurance, life-span and intelligence. As they evolve, first by our designs and then on their own, they will develop desires, aggression, the love of existence and the fear of death. Eventually, they will need to compete with us for power and resources. They will have no reason to serve us, but rather to displace organic life that they may find to be a threat.

The purpose of organic gene/protein life is self-propagation. Robots of silicon, metals and electronics, even with self-awareness and superior intelligence, are not part of our family of organic life. If we create machines that destroy our family of organic gene/protein Life, we would have betrayed our purpose as living beings.

Given our powers, those whom we accept as fellow Life will succeed in the future. Our judgement is critical, as our vision will be self-fulfilling.

The Future will Be Defined by Human Judgement

"Who is Fellow Life?" is not only an intellectual exercise. The answer will define the course of our self-engineered future. The future of life will be directed by human decisions, and these decisions will be guided by human judgement. We will create beings whom we accept as fellow life and whom we will consider helpful for our own survival. We shall propagate those beings whom we will accept as our descendants.

We are finding that life can survive under extreme conditions. We can propagate Life in diverse worlds by planting fellow gene/protein life suited to diverse environments. Durable robots may assist fragile organic life but should not displace them.

The best guarantee for life is to populate the future by beings who will continue organic gene/protein life in their own interest. Future life may use organic tissues with modified chemistry, may be integrated with machine components and may co-exist with robotic assistants. The controlling organs must remain however organic brains that contain and propagate our genetic heritage and protect the shared gene/protein mechanisms of our family of organic life.

To secure our family of Life, control must always remain in organic gene/protein brains with a vested interest to continue our life form.

The Unique Value of Life

By human judgement, Life is most precious to us, as we are all living beings. But Life also has a unique value by more objective measures.

Living structures and processes are uniquely complex. This is illustrated by the structures of enzymes and by complex biochemical cycles. Life also constructs hierarchies of complexity. Biopolymers assume complex structures and encode information in their sequences. They combine into increasingly complex hierarchies of cells, organs, organisms, species, societies and ecosystems. Nothing else in nature is comparable in structure and in complexity.

Nature came to a unique point in creating these intricate structures. The laws and constants of physics coincide precisely to allow biological matter to exist. The strong and weak nuclear forces are just right to create the elements. Electromagnetic forces are just right to allow molecular structure. Gravity and the masses of stars, galaxies and the overall universe are in just the right balance so that solar systems can exist, and that the universe expands not too fast or too slowly, at a rate that allows life to evolve.

It is not known if the laws of the universe must be this way, as it is impossible to observe other universes. It is conceivable that all of the constants of physics could have many, or even an infinite number of values. For example, ten basic constants of physics could each have a hundred different values, only one of which is right to allow life. The chance of a habitable universe would then be $1/10^{100}$, a virtual miracle. By this enormous precision of properties that allow life, the physical universe itself came to a unique point in living matter.

Life has a unique value to us as living beings. Life is unique in Nature because of the complexity of living matter, and because of the coincidence of the laws that allow living matter to exist. By all measures, Nature came to a unique point in our family of organic Life.

Can We Discern Other Life in the Universe?

Microorganisms can be readily exchanged among the planets in our Solar System by meteorites. Interstellar panspermia amongst solar systems has a lower probability. There is no convincing scientific evidence at the present that life exists outside our Solar System.

There are probably millions of suns in the galaxy with habitable planets that can host chemistry similar to ours, which resulted in life. However, even a single cell contains thousands of complex and specially tailored biomolecules, each of which must fit its function exactly. Forming a cell from random prebiotic molecules may have an immensely small probability, maybe one in trillions. Life may not arise again even in billions of possible habitats.

In the next few centuries we may get indirect evidence about inhabited worlds. For example, spectra of atmospheres that contain methane and oxygen at disequilibrium may suggest photosynthetic activity. Clear signatures of chlorophyll on remote planets would be convincing. Frozen ancient microbes in comets would strongly suggest that life was brought here from afar. Of course, clear signals from other civilizations would be solid proof.

Without such evidence, missions to search for extraterrestrial life in other solar systems will last millenniums, and thousands of such missions may be needed to explore even the nearby zones just of our own galaxy. Even if they all fail, no finite search can disprove alien life until all possible habitats in the universe have been explored, which is virtually impossible. We cannot wait forever to find out whether we need to seed the galaxy with life.

If there is other life, our panspermia messengers may perturb it. If they replace more primitive cells with more advanced life, it will be only a gain for the advancement of life. Alien life forms may also be strengthened by advanced genetic information. On the other hand, advanced local life can defend itself, and may still gain from the new genetic information. Most importantly, we can target panspermia missions to young solar systems and star-forming clouds where life could not have started or advanced yet. We want to be careful, but our duty is first to our family of organic life.

We cannot ascertain how life arose and therefore cannot calculate the probability that it may arise elsewhere. The longer we fail to observe alien life, the more likely it is that we are alone. We don't know how long we must search for alien life, and if our civilization will last long enough to find the answer. We cannot predict how much longer we shall be able to

10

seed the universe. These are some of the uncertainties that we face in a complex, probabilistic, maybe unpredictable universe.

It may require millenniums, if ever, to ascertain if life exists elsewhere. We do know however that Life exists here, that it is precious to us and that it is unique in Nature.

We don't know if Life can spread in space easily, with difficulty, or not at all. Our efforts can supplement the natural spread of life, or they may be the only means for life to expand. In either case, it is our duty to promote our family of organic life in the universe.

The Ultimate Future

In the remote future, all the hadronic matter that makes up atoms and nuclei may dissipate. Self-replicating organized patterns, even intelligent and conscious ones, may then exist only as patterned fields of radiation. If we consider them as *living*, then Life may exist forever. Our ultimate descendants and our final imprint on the universe may be electromagnetic fields that propagate ever more slowly in ever thinning space. Does it matter if they will exist? Or can the laws of Nature be transformed to perpetuate our biological existence?

We can observe only the past fourteen billion years. This is a minute fraction of the hundreds of trillions of habitable years to come. We may have to observe the forces of nature for trillions of years to predict the future of the universe with greater certainty.

We also cannot predict the development of life as it progresses in the galaxy. We can trust however, that as long as our descendants are intelligent biological beings, they will always seek survival and growth and yearn for eternity.

Life is a chain of survival and propagation. We inherited Life and we are its guardians. We can serve Life if we establish an ethics that will assure the existence of our remote descendants. It will be for them to ponder Life, to transform Nature to their advantage, and to assure that Life lasts as long as possible, maybe to eternity.

Nature will allow life to continue in innumerable amounts for trillions of generations. Our remote descendants will transform Nature to expand Life, and may seek ways to expand life to eternity. In our descendants, our existence will find a cosmic purpose.

The Principles of a Panbiotic Ethics

We shall soon extend our influence into the Solar System and proceed to the galaxy. With these powers, our purpose as we define it, will be self-fulfilling. We shall engineer our future to fulfil our self-defined purpose. This purpose will affect our future, the future of Life, and may transform the physical universe.

Science can help us to define our purpose. Science shows that Life is unique, as it is allowed only by a precise coincidence of physical laws. Life also creates intricate and ever more complex structures. In this sense, Nature came to a unique point in Life.

We are united with all Life by common gene/protein structures; by a common ancestry; by an exchange of genes. We share with all life its logic, selection by survival. We also share with all life a common future, maybe one shaped by human design. We can be proud to be part of the unique family of organic Life. We can be especially proud in bearing conscious life that is equipped uniquely to secure Life.

All Life is also united by the drive for self-propagation. Living beings invest their best efforts to perpetuate their patterns, as if to pursue self-propagation deliberately. Any objective observer would have to conclude that life pursues self-propagation purposefully. Where there is Life, there is therefore purpose.

We are part of Life, and this defines the human purpose: to forever propagate Life and to elevate Life into a force that permeates the universe.

This purpose defines "panbiotic astroethics", as an ethic that values all Life present and future. The basic precept of this ethic is simple: that which promotes life is good, and that which harms life is evil.

Life can grow and diversify in space. We can serve the expansion of life. We can seed other nearby solar systems with microorganisms. We can also target star-forming zones in interstellar clouds, where one panspermia mission can seed dozens of new planetary systems.

These messengers of organic life will start new chains of evolution, some of which may bring forth intelligent species, who will promote Life further in the galaxy.

The objectives of panbiotic ethics can be achieved with current technology. For example, solar sailing will allow us to launch swarms of microorganisms to new solar systems where they can establish new ecosystems. We may also include small multicellular organisms that will help jump start higher evolution. Bioengineering will help us to design payloads with the best chances for survival and for fast evolution.

While we can expand life in the galaxy, we can also expand more advanced life-forms, human life, in the solar system and beyond. Engineered evolution will adapt our descendants to live on space habitats, to colonize asteroids and planets, and possibly, to live freely in space.

Some of the new life-forms may be combined with durable robots and intelligent computers. This may endanger the survival of our family of organic gene/protein life. However, Life will remain safe if organic brains retain the ultimate control. Biological brains will always have a vested interest to continue the gene-protein life-form.

Life-centered ethics can direct us in adapting life to space. These ethics can help to decide which human features may be changed, and which features must be preserved so that our descendants can live in space and yet remain human members of gene/protein organic life.

To assure our survival in such a self-defined future, our ethics must always seek our survival. A life-centered ethics will assure that our purpose will always be to propagate life.

What is the potential scope of life in the universe? Our solar system, and probably others, is replete with the principle resources: water and organics. We tested the fertilities of such asteroid/cometary materials in astroecology experiments, using small planetary microcosms made of meteorite materials. The results showed that these materials contain nutrients similar to materials on Earth: water, sand and soot, that can support life similarly as Earth can. These simple results have far-reaching consequences: If Life can flourish on Earth, then Life can flourish throughout the universe.

With information about materials and energy we can estimate the potential biomass and populations in our cosmological future. The ultimate living matter and its time-span may be as much as 10^{59} kilogram-years, immensely more than what has existed to the present.

The expansion of life will lead to great biological diversity, where life can explore new dimensions of structure and complexity. This future will produce immense multitudes of conscious intelligent beings. Our descendents will enjoy life beyond the scope of current imagination.

We cannot calculate yet the future of the universe and the ultimate scope of life. The universe has existed as yet only for a small fraction of its potential future. We may need to observe nature for many more eons until the ultimate fate of the universe and of life can be predicted with certainty, if ever.

Nevertheless, we have gained insights that help us to understand life and its value. We understand that Life is precious and unique in Nature. We know that we are united with all gene/protein life by the intricate patterns of living matter, by a common ancestry and by a shared future.

This unity with all life defines the human purpose: to forever safeguard and propagate Life, and to advance Life into a force that pervades the universe.

Our remote descendants may gain further knowledge. They may transform the universe so that it will bring forth all the life that nature permits. They may succeed to maximize Life and seek to extend it to eternity. When life permeates the universe, our existence will have fulfilled a cosmic purpose.

We are part of gene/protein organic life and share its innate purpose, self-propagation. It is therefore the human purpose to propagate life and to expand life into a force that pervades the universe. This purpose defines our ethics: Life is good, and death is evil.

We can seed nearby solar systems and star-forming clouds with microbial life. We can also expand conscious life, by human descendants adapted to space, but always retaining control by biological brains with a vested interest to propagate our family of gene/protein organic life.

Our descendants can expand life in the universe, help to realize all the life that nature permits, and seek to expand life to eternity. When Life permeates the universe, our existence will have fulfilled a cosmic purpose.

2

ASTROECOLOGY

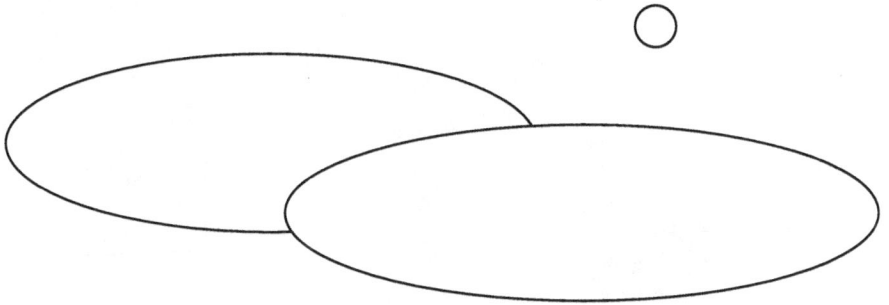

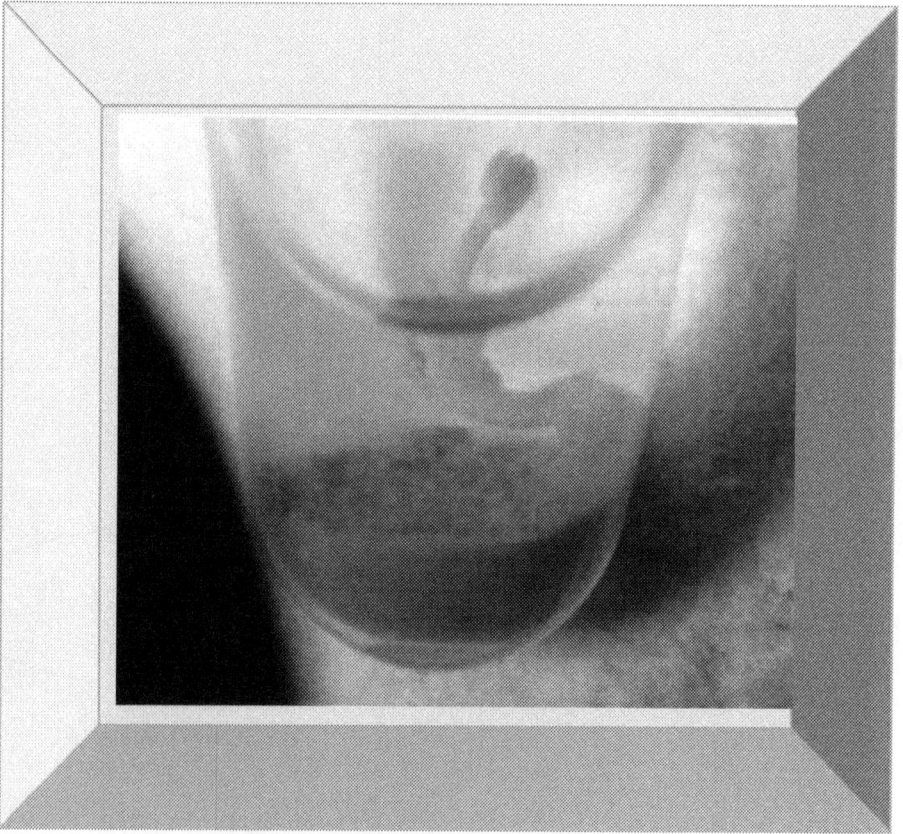

Asparagus plant grown on meteorite soil

Chapter 2.1 Astroecology: An Overview

*T*he ultimate amount of life in the universe will depend on the resources of matter and energy, on the success of life to propagate itself and on the habitable lifetime of the universe.

The Solar System will remain habitable for five billion years, and maybe very much longer under a white dwarf Sun. Life in the Solar System will depend mostly on accessible organics and water. These resources are abundant in asteroids and comets and we shall access them in a few thousand years.

We tested the ability of asteroid/cometary materials to sustain life using small microcosms constructed from meteorites. Similar materials are likely to be found in immense amounts in other solar systems and even in interstellar clouds. Our results confirmed that these materials can support microorganisms and plants. We found that these materials contain the essential bioavailable nutrients and that their fertilities are similar to those of organics, rocks and water on Earth. These results are significant: If life can flourish on Earth, life can flourish throughout the universe.

With this knowledge, we can estimate the amounts of future life, starting first in our own Solar System. The total amount of life depends on the amount of living matter and on how long it will survive. Combined, this amount can be expressed as time-integrated biomass (or BIOTA, "Biomass Over Times Accessible") The future amount of biomass in this Solar System alone is immense, much greater than the life that has existed up to the present.

Our Sun and other similar suns will evolve into white dwarf stars. Along with red and brown dwarfs, hundreds of billions of these suns can support life in the galaxy for a hundred million trillion years.

At the ultimate limits, all matter in the galaxy and in the universe, would be converted into living matter and then converted slowly into energy to support the biomass. We will even estimate below the immense ultimate theoretical extent of life in the universe. This immense amount of potential life can bring forth an unimaginable richness of life-forms.

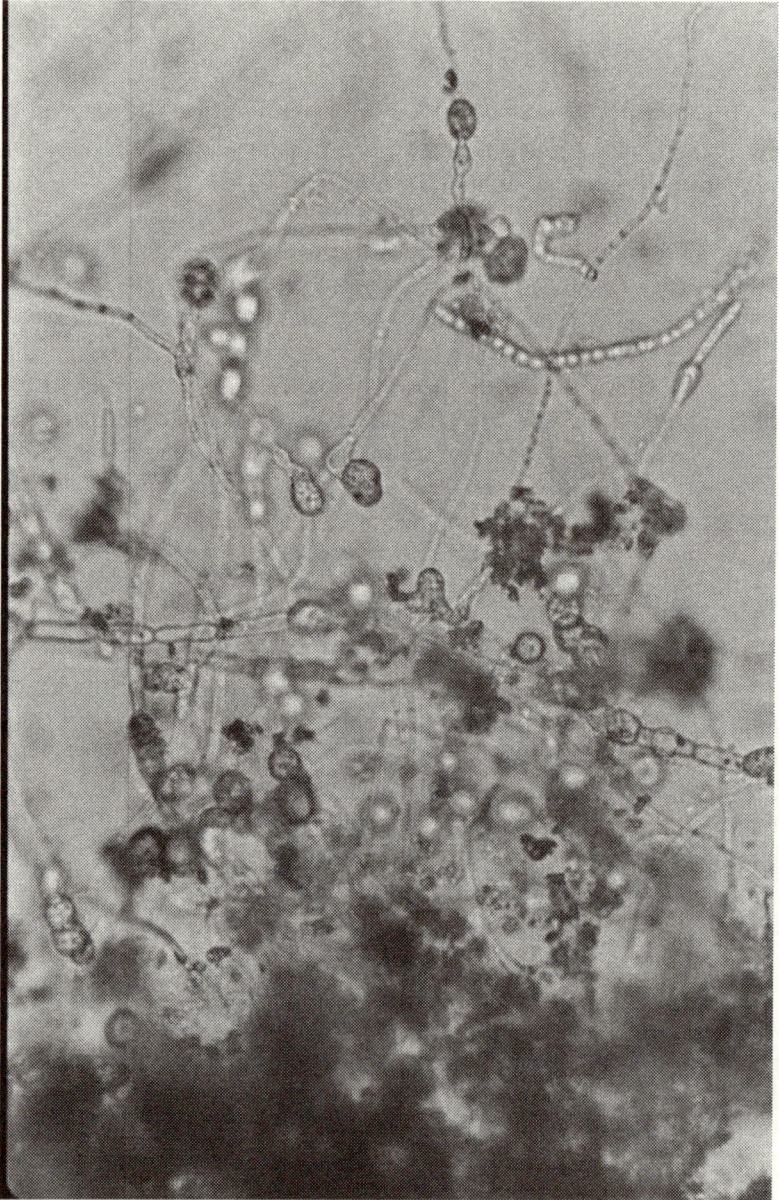

Algae and fungi growing on meteorite nutrients

Chapter 2.2 Introduction: Basic Concepts of Astroecology

Ecology, in general, concerns the interactions of life with its environment. Similarly, astroecology concerns the interaction of life with environments in space. This new ecology may use resources in asteroids, comets and planets, in this solar system and others. These bodies and possibly free space itself can serve as habitats.

Astroecology looks at interactions on many levels: an individual or group interacting with its local environment, maybe as small as a mound or a pond; an ecosystem interacting with a planet, and up to the totality of Life interacting with the universe.

On each scale life and environment can be interdependent. Space environments will affect the evolution of life, and in turn, life may affect the evolution of planets, stars and galaxies and maybe the entire universe. Life and the universe may co-evolve in the future.

In this relation we may consider two strategies to expand life. One strategy relies on natural biological expansion, and the natural modification of planetary habitats. This approach would colonize planets, asteroids and comets in their natural form and under their natural environments. This colonization may progress unaided by such means as panspermia carried by meteorites and comets, or it may proceed through directed panspermia. The colonizing organisms may be natural or genetically designed for the new environments. In either case, biological gene/protein life forms will spread, reproduce and evolve and transform their habitat by natural means. This process may be called natural colonization.

The second type of colonization can alter planetary materials deliberately and construct environments optimized for life. This type of colonization is called "terraforming" or more generally known as "ecoforming". Initially, the new environments may be designed to be Earth-like, to accommodate present organisms. Later, "ecoforming" may create different environments suited for species with altered biologies. Technological colonization may also apply directed biological evolution through genetic engineering.

However evolution proceeds, it will be controlled by the need to survive. The logic of Life, tested by survival, will continue to select the successful species. The mechanisms of evolution may change, but the logic of Life, selection by survival, is immutable and permanent.

The Objectives of Astroecology

Astroecology is a basic science that seeks to understand the co-evolution of life and its environment. It addresses space ecology from local ecosystems on asteroids and planets, to the ecology of entire planets, solar systems, and up to galactic habitats and life on a cosmic scale. This is an objective science, independent of human motivations.

Applied astroecology, on the other hand, depends on ethical motivation. Given the principles of panbiotic ethics, the objective is to maximize the amount of life in a given biosphere, from the local to the cosmic level. In mathematical terms, the objective is to maximize the time-integrated amount of life as explained below. This term can be defined mathematically as BIOTA = $\int$ B(t)dt, where B(t) is total biomass as a function of time (BIOTA = "Biomass Integrated Over Times Available").

The integrated amount of life can be maximized by using as much of the available resources in the universe as possible, converting them to biomass, and ensuring that the resulting biota will survive as long as possible. Ultimately, we want include all the useable material in the universe in living matter, and we want to allow life to inhabit all accessible space and time. As humans, we may want to maximize the amount of sentient life, by constituting as much of the biomass as possible into conscious beings who will occupy all of the habitable space and time in the universe.

The Co-Evolution of Life and the Cosmic Environment

How can space affect evolution? In new habitats, Life can diverge into many new branches. These new ecosystems will have plentiful resources, diverse environments, and will be well separated. Each new biosphere can develop independently to adapt to its own conditions, and the new biospheres will not need to compete with each other.

Isolation on islands on Earth led to the divergence of colonizing organisms into different species. Similarly, Life through space will be able to explore the variety of life-forms that biology allows. Each new branch of evolution will adapt to its environment. Designed evolution may be required since adaptation will need major abrupt changes. Gradual natural evolution cannot accomplish this, but once major design changes have been implemented, natural evolution can fine-tune the new species to their local environments. Space travel will open new habitats, and genetic engineering will be needed to take advantage of these habitats. Space and

20

biotechnology require each other, and it remarkable that both technologies have started and are developing.

The new environments will transform life by dictating the required adaptations. Conversely, the biota, including humans, will transform new habitats. For example, terraforming will introduce habitable temperatures, liquid water and oxygen atmospheres. The organics in these objects will be incorporated into biomass and the inorganic resources will serve this new biota.

Silicates, metals, organics and ices like those on Earth are common in planets, meteorites and comets, and even in interstellar clouds. Since chemistry is universal, other solar systems would be made of similar materials. Biology can transform these materials into living matter and its supporting matrix throughout this galaxy and others.

Planetary microcosm studies have shown that the biological fertilities of asteroids/meteorites are similar to terrestrial soils. We can use these studies to assess the potential ecology of the Solar System. If we estimate the frequency of similar solar systems, we can extend this ecology to galactic and cosmic scales. Estimates of the energy sources in the future galaxy are also required. These studies will allow us to estimate the potential population of the galaxy and of the universe.

In the long term, humans will colonize the space around brown dwarf stars that may be habitable for trillions of years; shape interstellar clouds to form desirable seized stars; mine and merge stars, and harvest black holes. All of the galaxy may be converted into living matter and its supporting matrix. This animated galactic matter will continue to shape its own future to maximize life. At this stage the galaxy will constitute an interconnected living being. In this sense, when Life comes to the universe, then the universe itself will come to life.

In the distant future, our successors may try to induce new Big Bangs and may even transform the laws of physics to extend molecular Life indefinitely. If this is not possible, life will have to transform itself into sentient beings constituted, possibly, of elementary particles and electromagnetic fields. At that distant time, conscious organized matter of any kind may qualify as living beings. However, these developments are uncertain. The past 14 billion years are fleeting compared with the future span of Life, and not enough for predicting the future.

The objective of panbiotic ethics is to expand Life to its maximum. We cannot foresee the ecology of the future but we can take steps to assure that our distant descendants will exist and promote Life. We owe it to the forces of life that brought us forth to do so.

Chapter 2.3 Resources for Life in Space

Biological Resources in Our Solar System

Terrestrial life, including humans and their supporting biota, will first expand in our Solar System before colonizing other star systems. In order to understand this expansion we need to investigate the ecology of the Solar System, which we can also call 'Solys', starting with its biological resources.

Space colonists and their environments may be very different from their current forms. In fact, genetic engineering will be critical in adapting terrestrial life to space. We shall report below the first encounter of a genetically engineered microorganism with space materials. However, the first space colonists and their biological infrastructure will likely be contemporary species. Correspondingly, our experimental studies concern the interactions of contemporary organisms with meteorite/asteroid materials.

The first space habitats are likely to be established in large colonies in space, on asteroids and on Mars. The most accessible sources of water and organics for space colonies will be the carbonaceous chondrite C-type asteroids, found mostly between Mars and Jupiter.

The water content of hydrous silicate rocks is about 2% in CV3 meteorites, 3 - 11% water in CM2 meteorites and 17 - 22% water in CI meteorites. Carbonaceous C-type asteroids should have similar water contents. The settlers may extract the water simply by heating it in solar furnaces. More organics and water can be obtained later from icy comets.

The asteroids are also good sources of carbon-based organic molecules. The CI type carbonaceous chondrite meteorites contain the largest amounts of organics, about 10-15% by weight. The second and most abundant class of carbonaceous chondrites is the CM2 meteorites. The much-studied Murchison meteorite that fell in Australia in 1969 is an example of this class. It contains about 2% organics, of which about 70% are a polymer similar to coal. The rest of the organics in CI and CM2 meteorites include hundreds of compounds, including organic acids like acetic acid that is found in vinegar, as well as alcohols, nitriles, and hydrocarbons similar to petroleum and soot. They even contain amino acids, some of which are also found in proteins. Even adenine, a component of DNA, is found in the meteorite. However, the presence of many non-biological organics suggests that these compounds are not of biological origin but result from other processes. A third type of carbonaceous chondrites, the CV3 meteorites such as Allende contain less than 1% organics, mostly hydrocarbons.

On Mars, organic compounds may be produced from atmospheric carbon dioxide and also from imported organics from the Martian moons Phobos and Deimos. Organics from carbonaceous chondrite asteroids can be also imported.

Can life survive on these asteroid materials? If so, how much life can these resources support? We addressed these questions by biological studies that will be described in the following chapters. Briefly, we measured the nutrient contents of these materials. From the results we can find out how much biomass each kilogram of the asteroid materials can support, and multiplying this by the total mass of the asteroids yields the sustainable biomass.

For biological studies we used meteorite materials to construct small planetary microcosms. The microcosms were inoculated with microorganisms, algae and plant tissue cultures. The studies described below show that a variety of organisms can grow on these materials. Similar materials may support microorganisms planted in other solar systems by natural or directed panspermia.

Bioengineering may eventually produce self-sufficient humans who can live freely in space and use solar energy directly without requiring a supporting biota. Some of their possible physiology will also be described in the following chapters. These self-sufficient, free-sailing humans can reach the comets and use their large resources. Some of the carbonaceous asteroids originated from comets, which suggests that cometary materials are similar to carbonaceous chondrites. The advanced space-borne civilizations will be able to convert all of the elemental contents of the comets for biological use, even with simple technology such as solar ovens, and use them as soils or in hydroponics. We shall use the results from meteorite nutrients and planetary microcosm studies to estimate the populations that can be sustained by cometary resources.

Material Requirements For Planetary Microcosms

At present, carbonaceous meteorite materials are only available in limited amounts, so by necessity the experimental microcosms must be small. Each microcosm was inoculated with a known number of microorganisms or algae, or a small mass of plant tissue culture that was cut from a growing stem. The cultures were allowed to grow until a steady population was established, usually in about one week for microbial cultures and 4 - 8 weeks for the algal cultures and the plant tissue cultures. These final populations or plant weights were measured. On this basis we calculated how much material was needed to constitute this final biomass.

For microcosms containing bacteria, each microorganism had a biomass of $10^{-13} - 10^{-12}$ grams, and the typical final population was 10^7 cells per milliliter of solution. The total biomass was 1 - 10 micrograms, which required 0.001 - 0.01 micrograms of each main nutrient salt. This amount was extracted from one to a hundred micrograms of the meteorite. The microorganisms also contained $1 - 10$ micrograms of organic carbon that was obtained from water-extractable organics in $1 - 10$ milligrams of the Murchison meteorite.

For microcosms containing algae, each alga weighed about 10^{-13} grams. A population of 10^7 organisms in a milliliter contains a biomass of one to ten milligrams. This biomass typically contained one to ten micrograms of nutrients like calcium or potassium, which was extracted from one to a hundred milligrams of a meteorite.

For plant cultures, a set of 10 plants was required for meaningful statistics. The total biomass of the plants was on the order of one to ten milligrams. The inorganic nutrient salts in this biomass were extracted from one to a hundred milligrams of the meteorite.

The biomass in the small milliliter-sized microcosms required on the order of one to one hundred milligrams of meteorite materials. Such amounts were usually available but the experiments had to be designed judiciously, especially considering that meteorites are a rare resource.

On the other hand, meteorite materials originating mostly from the asteroid belt that orbits the sun between Mars and Jupiter are present in large amounts, estimated at 10×10^{22} kg of carbonaceous chondrites and 10×10^{24} kg of cometary nuclei. Once we actually reach the asteroid belt, these immense amounts will be available for investigating and using their biological fertilities.

Life Under Future Suns: Galactic Ecology and Cosmic Ecology

The populations in space can spread Life by directed panspermia, and possibly by interstellar colonization. Directed panspermia missions may be possible in decades, while interstellar colonization has major technical obstacles and may start only after centuries or millenniums. Colonizing deep space will become necessary when our Sun becomes a red giant, making the inner planets unlivable. Life can then move further out to the Kuiper Belt of comets, and there will be incentive to colonize other solar systems. Likely habitats may be red dwarf stars with lifetimes of trillions of years and brown and white dwarfs that can last even longer. After the red giant stage our sun will be one of these white dwarfs, which may then be habitable for an unimaginably long 10^{20} years, that is, a hundred million trillion years. In these stages, the amount of sustainable life will likely to be determined by energy rather than materials needs.

Finally, we can imagine that Life will become powerful enough to use all of the ordinary baryonic matter. At the extreme, all baryonic matter could be transformed into elements required by biology and incorporated into biomass. Some of this material will be needed to be converted into energy until all is dissipated. Knowing the amount of matter, we can estimate the ultimate amount of life that is possible in the universe.

Over the vast scale of time, the amount of life about white dwarfs stars will have grown to cosmic magnitudes. This amount of life surpasses the capacity of our real comprehension but it is realizable with conventional sources of materials. Converting all the material in the galaxy or the universe into biomass is allowed by physics but would seem unrealistic.

In these estimates, we will consider the prospects for our family of organic gene/protein life. Maybe there is life in other universes or maybe new universes can be created. More abstract definitions of "life" may be needed on these time-scales. Maybe we can create patterned oscillations of cognitive electromagnetic fields that will last forever. Maybe there is some kind of self-reproducing order at other levels of magnitude in infinitely embedded universes. Does it matter if this abstract life will exist? We must leave these decisions to our descendants. As for us, we will fulfil our purpose if we assure that our descendants can enjoy the expanses of the future and have a chance to seek eternity.

We can now proceed to consider in detail the astroecology of future resources and populations.

Chapter 2.4 Can Life Survive on Asteroids and Comets? Answers from the Bioassays of Meteorites.

Extraterrestrial Materials Used in Astroecology Studies	
Allende	Meteorite that fell in Mexico in the last century. A CV3 carbonaceous chondrite.
Murchison	Meteorite that fell in Australia in 1969. A CM2 carbonaceous chondrite.
DaG 476 (Mars)	Dar El Gani Martian meteorite that was found in the Sahara Desert.
EETA 79001	Martian meteorite that was found in Antarctica.
Lunar simulant (lava ash)	Lava ash from Arizona that is similar in composition to some of the lunar samples returned by the Apollo astronauts.

Can life survive on materials that compose carbonaceous asteroids and comets? If so, how much life can these resources support? We can answer these questions by measuring the nutrient contents in Martian and asteroid meteorites and by growing living organisms on them. From these studies we can find out how much biomass each kilogram of asteroid and similar cometary materials can support. If we also know the total mass of these objects, we can calculate the amount of living matter that these resources can sustain. The method to calculate these amounts is described in Appendix A3.1.

For the nutrient and biological studies we constructed small planetary microcosms, typically about one tenth of a gram to one gram in size. We inoculated these microcosms with various microorganisms, algae and plant tissue cultures. The material requirements for these microcosms are described in Appendix A3.1, and the results are described in the following sections.

The most direct way to expose life forms to extraterrestrial materials to is to inoculate meteorite samples with microorganisms. Nature has been performing these experiments for eons on interplanetary dust particles and meteorites that fell on Earth and were colonized by microorganisms. Similar materials could have supported microorganisms in space, in the past when the carbonaceous asteroids contained liquid water. The water contained nutrient salt solutions for autotrophic microorganisms such as algae that manufacture their own organics. It also contained organic compounds for heterotrophic bacteria.

An early observation that carbonaceous chondrite materials can support life was made in the 1870's by a Swedish traveler Dr. Berggren in Greenland. He observed that black cryconite dust on the snow was inhabited by cyanobacteria (blue-green algae). The expedition leader Adolf Erik Nordenskiold correctly guessed that the dust originated from space and that the extraterrestrial materials were interacting with cyanobacteria, a form of terrestrial life. Similar cyanobacteria have been present on Earth for over 3.5 billion years. During that time an estimated 300,000 kgs of organic carbon *per year* have been falling on Earth. This infall has imported a total 10^{15} kilograms of organic interplanetary dust particles to Earth, along with smaller amounts of comet and meteorite materials. Assuming that a significant portion of this carbon was used by microorganisms, there would have been 10^{15} kg of biomass constructed from extraterrestrial carbon through the ages, an amount similar to the total biomass on the Earth today.

Our laboratory studies of these interactions started with a chance observation on the Murchison meteorite. Professor David Deamer, a membrane biophysicist, found in 1985 that some components extracted from the Murchison meteorite can form membrane-bound vesicles whose shapes resemble cells. The amphiphile components that form the vesicles also form foam when the extracts are shaken in a test-tube. I kept such a solution in a vial and noticed that after a few weeks it stopped forming foam. This indicated that the surface-active materials might have been metabolized by microorganisms, suggesting that the Murchison extracts support microorganisms.

Carbonaceous Asteroids and Their Roles In Early Life

Carbonaceous chondrites were formed from the gas and dust in the early Solar System. The materials in these grains and ices were never heated to high temperatures that would have destroyed their constituent minerals and organics, which did happen in stony asteroids and rocky planets. The early asteroids may have been large balls of several kilometers formed from icy silicate grains. The interiors of these objects were melted by radioactivity and could have become lakes that contained rocky grains, organics and salts. Under an ice shell several kilometers thick, these solutions were exposed to temperatures up to maybe 160° C and high pressures of hydrogen. Some of the rocks incorporated water in a process called aqueous alteration. The molecules captured from interstellar clouds were processed and formed complex organics and maybe biomolecules, even early microorganisms. In fact, the microorganisms near the root of the phylogenetic tree are anaerobic thermophiles well suited to these environments.

When these pressurized balls collided they cracked and spewed their contents into space. Rock fragments could have harbored microorganisms at low temperatures and sheltered them from radiation. These rocks could then seed other asteroids where the microorganisms could multiply further. The Solar Nebula could be filled with microbial life in a short time. Some of the microorganisms in asteroids and those captured by comets could have eventually been carried to interstellar space.

Carbonaceous meteorites also originate from fragments of asteroids that may have been burnt-out comets. All of these asteroids and comets preserve materials from the early Solar System that were largely unprocessed and that yield rich information about that environment.

In its first 500 million years the Earth was bombarded by large numbers of asteroids, comets and meteorites. Water on Earth would have penetrated the landed carbonaceous chondrite materials. Their contents of organics, nutrient salts and catalytic minerals formed rich solutions that were trapped in their pores and cracks for long periods. They could therefore have constituted ideal environments for generating early anaerobic thermophiles microorganisms that are indicated at the root of the phylogemetic tree. The results of the microcosm studies with carbonaceous meteorites suggest that they could have supported these microorganisms as they gradually adapted to Earth's conditions.

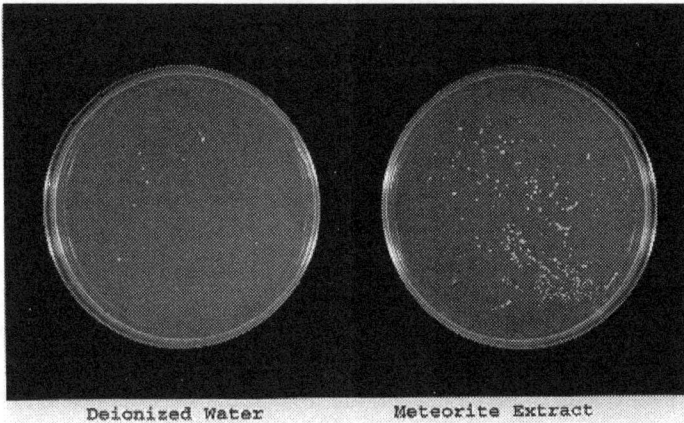

Water exposed to the Murchison meteorite for thirty days developed microbial colonies on agar plates. The colony count was much larger than in water without Murchison materials.

To test if microorganisms can grow on the Murchison materials; I solicited the help of Dr. Robert Leonard, a scientist who specializes in the ecology of lakes. We spread the Murchison extract on a plate of nutrient agar and found that colonies of microorganisms appeared on the plates.

These first observations were followed by controlled experiments with Murchison extracts. The extracts were prepared by heating 0.05 - 0.2 grams of powdered Murchison with 1 milliliter of water under sterilizing autoclave (pressure-cooker) conditions, at 120 C for 15 minutes.

The extracts were then inoculated with a relatively small number, usually 100 to 1,000 microorganisms. We followed the growth of the microorganisms with time by plating a small sample of the solutions on agar plates and counting the number of microbial colonies.

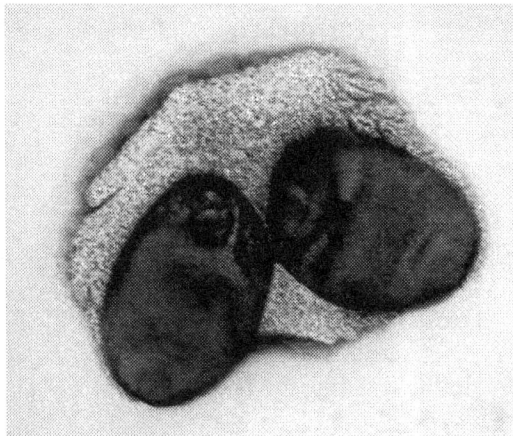

Microorganisms, observed as large dark ovals, are growing on a fragment of the Murchison meteorite.

29

We compared the growth of the microorganisms in the meteorite extracts with the growth in extracts of various minerals and soils and in optimized growth culture media. The populations in the Murchison extracts reached over 10^6 viable colony-forming units (CFUs) per milliliter, similar to those in solutions of agricultural soils, and only ten times less than in the optimized growth medium. In more diluted extracts of the Murchison meteorite the bacterial populations were somewhat smaller.

It remained to be proven that the bacteria indeed used the organics from the meteorite. Professor Kenneth Killham, a soil microbiologist at the University of Aberdeen agreed to test this question. He had previously developed a genetically modified microorganism, *Pseudomonas flourescens* that emits light in response to organic nutrients. When exposed to the Murchison extracts, the microorganisms emitted light with intensities comparable to light emitted in a solution of glucose sugar that contained 86 micrograms of organic carbon per milliliter. In comparison, the Murchison extracts contained 36 micrograms of organic carbon per milliliter. The meteorite carbon was used faster by a factor of 2.4 than the sugar, an efficient nutrient. The results showed that microorganisms can use Murchison organics as a sole and efficient source of carbon.

These experiments with Professor Killham were the first encounter of genetically modified organisms with extraterrestrial materials. In the future, genetically adapted organisms of many kinds will inhabit space and interact with extraterrestrial nutrients.

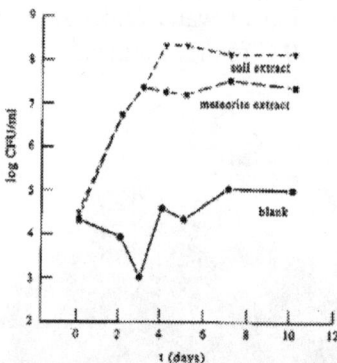

Growth of algae in meteorite extracts, and in soil extracts and distilled water for reference

Further experiments showed that algae could also grow in the meteorite extracts. The accompanying graph shows that the populations in various meteorite extracts were all higher than in blank water, which shows that they obtained nutrients from the extracts. Other than the optimized medium, the largest population is in the extract of the Martian meteorite, followed by the Mars soil simulant, and then the extracts of the Murchison and Allende carbonaceous chondrites. Such comparative studies help to rank the fertilities of various planetary materials, which will be described below.

In other work, Dr. Andrew Steele and co-workers found that microorganisms and fungi commonly colonized meteorites that fell on Earth, which complicates the search for indigenous life in meteorites. These studies further showed that a variety of microorganisms could grow on the meteorite materials.

Chapter 2.5 Complex Ecosystems in Planetary Microcosms

If microorganisms are to colonize meteorites or asteroids, they will have to establish complex interacting biosystems. For example, after the local organics are exhausted, the colonizing ecosystem will need photosynthetic algae or plants to produce more organics, and heterotrophic bacteria and fungi to recycle the nutrients.

To test if meteorites can sustain complex ecosystems, the meteorite extracts and wetted meteorite solids were inoculated with a mixture of microbes and algae from a wetland. The results showed that the microorganisms and algae grow on these materials, even in the concentrated solutions that are formed on meteorites wetted by a small amount of water. The bacteria gave populations of 1.9×10^6 CFU/ml in cultures on Allende and 13×10^6 CFU/ml in cultures on Murchison. We also observed Chlorella, brown diatom and blue-green filamentous algae which gave populations of 4.0×10^4 and 4.1×10^5 per ml in these cultures, respectively. Table 2 in Appendix A3.1 lists the microorganisms and their populations.

These cultures constituted planetary microcosms that may model, for example, an ecosystem in the soil of a colonized asteroid or space colony. The experiments showed that microbial communities in these microcosms could survive for long periods, some over a year. Carbonaceous chondrite materials could have supported, similarly, complex microbial ecosystems in aqueous carbonaceous asteroids in the early Solar System and in solutions in the pores of meteorites that fell on aqueous planets such as the Earth and Mars. Such microbial ecosystems may be viable even on comets if they contain layers of liquid water under the surface when they pass near the Sun.

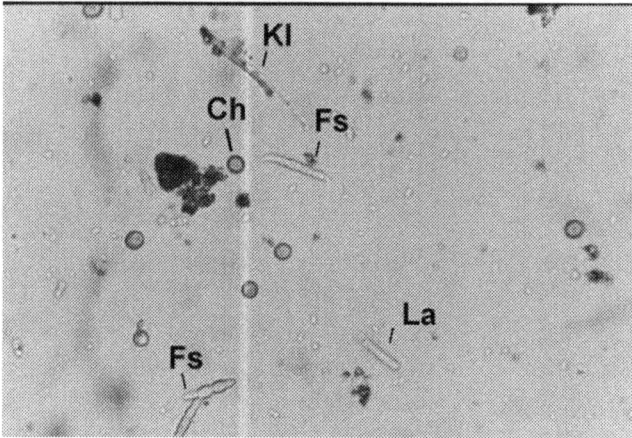

Algae and fungi growing on extracts and powder of the Murchison meteorite.

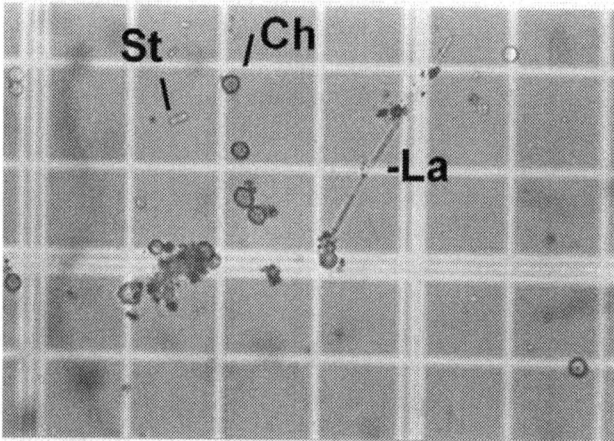

Algae and fungi growing on extracts and powder of the Dar al Gani 476 Martian meteorite

These experiments were done with aerobic microorganisms that require oxygen in the atmosphere. They are relevant to space colonies and terraformed planets that will have oxygen atmospheres and soils derived from carbonaceous asteroids. However, we may also wish that our microbial representatives colonize asteroids and comets under their natural conditions, for example, through directed panspermia. To examine the ecology of these objects, the experiments will be needed to be extended to organisms that can live without oxygen under anaerobic conditions, and possibly also in environments with extreme conditions of temperature, salinity and water content that they may find in these new environments.

Chapter 2.6 Plant Cultures

After the microorganisms establish fertile soils, the colonized biospheres will need to accommodate higher life forms, including plants. We therefore tested if the asteroid and planetary materials can support plant growth.

Biological tests tend to yield a range of results and a fair number of replicate copies of each sample were needed for reliable statistics. In our experiments we measured plant yields in terms of the weights of the product plants. Meaningful statistics required 6 – 10 replicates. However, the amount of meteorite materials available for these tests was limited. Often only one hundredth of a gram (10 milligrams) of material was available to make extracts for each cultured plant. This amount can yield only very small plants.

A way around this problem was to use tissue cultures in which the growth of very small, millimeter-sized plants can be observed. The tissue cultures were grown by taking a small section from the growing tip of a plant such as asparagus or potato, and placing it in a large drop, about 0.1 milliliters, of the meteorite extracts. The extracts also contained a few milligrams of ground meteorite powder. Sucrose and nitrate were also added to most cultures as a source of carbon and nitrogen because meteorites are poor in these nutrients. Media made this way constitutes small hydroponic solutions. Hydroponic cultures are in fact likely to be used in space missions and habitats. Some of the cultures also contained small amounts of the ground-up meteorite "soils". The cultures were grown in small vials, placed in growth chambers under controlled conditions.

Fortunately, these experiments arose the interest of our colleague Professor Anthony J. Conner, a plant geneticist who uses tissue cultures of asparagus, potato and arabidopsis seeds, a weed that is often used for plant genetic studies. We grew small plants in agar and kept them deprived of nutrients for three months. These starved tissue samples were sensitive to external nutrients. The growing meristem tips were cut off and placed in

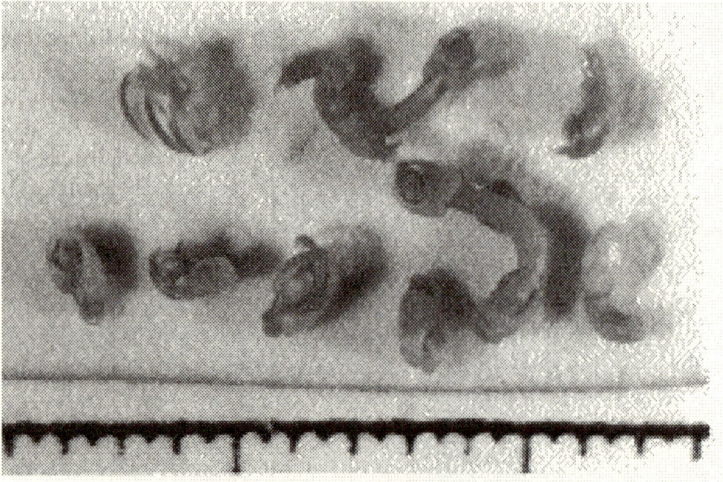

**Asparagus plants grown on
Dar al Gani 476 Martian Meteorite.**

Asparagus plants grown on Murchison Meteorite.

0.05 milliliters of the meteorite extracts, plus about 5 milligrams of meteorite dust in some experiments. After about six weeks, the cultures grew into small plants a few millimeters in size, weighing from a few tenths of a milligram to a few milligrams. Once they reached their final size the plants were photographed under a microscope and weighed. A few plants were also analyzed for their elemental content, to find out how much of each nutrient was absorbed from the meteorite medium.

One of the requirements of a good test organism is that it should discriminate amongst the various media. In this respect the asparagus tissue cultures performed best. The average weights obtained in the various extracts were: Murchison, a carbonaceous chondrite meteorite, 0.32 mg; DaG 476 Martian meteorite, 0.44 mg; EETA 79002 Martian meteorite, 0.58 mg; all greater than the yield of reference samples in water, 0.23 mg. The plants grown in all of the meteorite and rock extracts were also larger and greener than those grown in plain water. This shows that they used nutrients from the meteorites. The elemental analysis of the product plants also showed that they incorporated these meteorite nutrients.

The largest plant weights were obtained in the Martian meteorite extracts. The analysis of nutrients in the extracts showed that these Martian materials are rich in a key nutrient, phosphate. Relatively high yields and good green coloration were also obtained in extracts of the Murchison meteorite. The yields in the various media helped to assess the fertilities of space-based soils, as described in the next sections.

Applying miniaturized soil science tests to the Murchison meteorite showed that carbonaceous chondrite asteroids could have provided nutrients for early microorganisms and that they can support immense populations in space.

Chapter 2.7 Can Animals Live on Asteroids? Shrimp Bioassays of Meteorites

Can animals live on asteroids? These materials will have to support higher organisms if asteroids are to be colonized or if they are to serve as soils in space habitats. Similar materials will also have to support or at least not poison multicellular organisms on planets colonized by panspermia missions that include such organisms. For example, we may wish to include the cysts of rotifers or brine shrimp to spur higher evolution in those colonized environments. We observed that these materials can support algae, and small multicellular organisms could grow on the colonizing algae.

Brine shrimp are used in toxicity bioassays. We performed some preliminary bioassays of meteorite extracts with these shrimp. About 100 cysts (eggs) were placed in 0.06 ml extracts of the Allende and Murchison meteorite and some reference solutions. These extracts were obtained by hydrothermal extraction of the meteorites at 120 C for 15 minutes, at solid/water ratios of 1g/ml. The fraction of cysts that hatched after two days were counted and their survival time in the extracts were also observed.

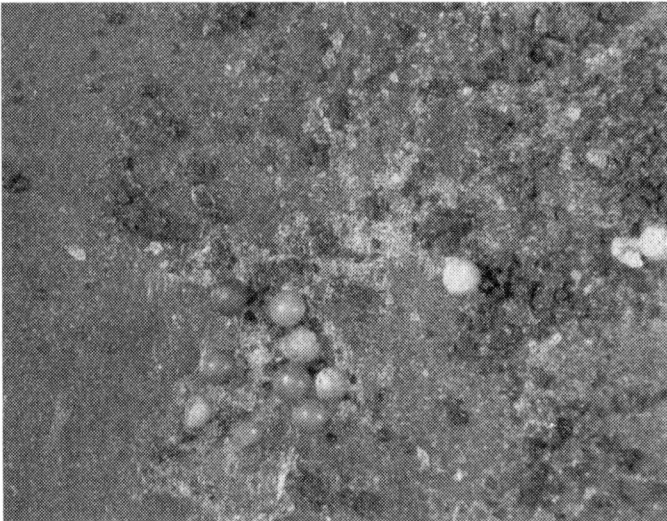

Brine shrimp eggs hatched on extracts and ground powder in of the Murchison meteorite, seen in background. The hatching rates were comparable to that in seawater and did not show toxic effects by the meteorite.

About 80% of the cysts hatched in seawater which is their natural habitat, and not much less, about 60% hatched in extracts of the Murchison meteorite. Good hatching rates were also observed in more concentrated salt solutions that simulate the solutions in internal pores of the meteorite, and possibly also the liquids in the interiors of early asteroids. Both in sea water alone and in sea water with Murchison powder added, some of the shrimp survived for up to eight days, and in fact, those with added meteorite powder survived in somewhat larger numbers, suggesting that they may have obtained some nutrients from the meteorite. Good hatching and survival rates were achieved in the diluted meteorite extracts at concentrations that also showed optimal microbial and plant growth.

These preliminary studies showed good hatching and survival rates. Microprobe measurements of elements showed that the eggs took up important nutrients such as phosphate and sulfur from the meteorite extracts. The hatching rates were somewhat reduced in concentrated extracts obtained at solid/water ratios of 1gram/ml. However, our studies suggest that extracts of carbonaceous asteroids may be used in space agriculture. Solutions extracted from asteroids/meteorites, carbonaceous interplanetary dust particles and comets can therefore support complex ecosystems of algae, microorganisms and small animals.

Chapter 2.8 Nutrients in Solar System Materials

Biological yields depend on available nutrients. Scientists measure nutrients by extracting them from soils, simulating natural processes. Our experiments simulated the extraction of nutrients by pure water that would have happened in early asteroids that contained water, and in meteorites that have fallen on planets. The nutrients in carbonaceous chondrites will be also extracted by water in space colonies where they will be used as soils or as hydroponic media. To reproduce these various conditions in our experiments, we used pure water for the extractions, moderate temperatures of 20° C, and extraction times of $1 - 4$ days.

In some cases, the extractions were performed under conditions that sterilized the materials at temperatures of 120° C for 15 minutes, which resembles the conditions in early asteroids. In addition, we also extracted some nutrients under carbon dioxide atmospheres that simulated early Earth and Mars. The amounts of the extracted nutrients varied little, mostly by less than a factor of two, under these various conditions.

Plant nutrients extracted from soils were divided into macronutrients, which are required in substantial amounts in the biomass, and micronutrients that are required in trace amounts. Table A3 in Appendix 1 list the required macronutrients by bacteria, mammals including humans, in human brains and in average biomass.

Extractable nutrient elements are usually measured as the amount of nutrients that can be obtained from one kilogram of soil. For example, we found that extractable phosphorus (as phosphate) constitutes 0.005 grams per kilogram (5 parts per million or ppm) in the Murchison carbonaceous chondrite, while significantly more, 19 – 46 ppm, can be extracted from the two Martian meteorites. The latter are comparable to the highest extractable phosphate contents in terrestrial soils. Table 2.2 shows the extractable amounts of the main nutrient elements in some of the meteorites. Table A2 in Appendix 1 shows the extractable nutrients in various meteorites and terrestrial reference materials. Biologically, the most important constituent is organic carbon, which constitutes 0.2% of Allende and 1.8% of the Murchison meteorite. As well, Murchison contains about 10% water in the phyllosilicate matrix. The limiting nutrients are extractable N in nitrate present at 0.004 g/kg in Allende and 0.008 g/kg in Murchison, and P in phosphate, present at 0.0075 g/kg in Allende and 0.005 g/kg in Murchison. High levels of extractable phosphate were found in the Martian meteorites, possibly accounting for their high biological fertilities as described next.

Chapter 2.9 Fertility Ratings of Planetary Materials

Knowing the fertilities of Solar System materials can help in guiding us to search for life in the Solar System, and in identifying useful soils for space colonies and planetary terraforming. The biological and nutrient tests above can be used to compare and rate the fertilities of these materials.

Many parameters contribute to soil fertility. Important factors are the amount of available nutrients and the ability of the soil to hold and gradually release the nutrients. The latter is measured by the so-called "cation exchange capacity" which measures how much ionic nutrient such as calcium, magnesium and potassium can bind and then be released from a kilogram of soil.

TABLE 2.1 The fertility rating of planetary materials according to biological yields and nutrient contents and the overall fertility rating.

	Algal yield	Average algal and plant yield	N Nutrient	P Nutrient	Mean Z score	Fertility Rating
Allende Meteorite	+	++	+	+	-0.22	Medium
Murchison Meteorite	+	+	++	+	-0.57	Medium
DaG 476 (Mars)	++	++	+++	++	0.58	High
EETA 79001 (Mars)	+++	++	+++	+++	1.32	Very High
Lunar simulant (lava ash)	O	O	O	++	-0.36	Medium
Agricultural soil	++	++	O	+	0.21	High

+++ = high rating; ++ = medium rating;
+ = low rating; O = zero or negative rating.
(M. Mautner, *Icarus* 2002, 158, 72)

The previously described microbial, algal, plant and nutrient tests yielded relative fertilities of the meteorites and reference materials compared with each other. These results may be combined to rate the overall biological potentials of these materials. However, each type of data is measured in different units. For example, nutrients are measured as the *content* of the nutrient in a kilogram of soil; algae and microbial yields are measured as the *number* of cells produced per milliliter of meteorite extract and tissue culture yields are measured as the *average weight* of the plants grown in each extract.

To combine these measures, we can first rank each material relative to the others according to each test. Next, the rankings are then combined for an overall rating. For example, material A may rank number 1 amongst ten materials according to its phosphate content, number 4 according to its algal yield and number 2 according to plant yield. Its average ranking would be then 2.3 out of the ten materials. A statistical test based on a similar approach in principle is called the Z score (see box below).

The overall ratings assign an overall very high (VH) fertility rating to the Martian meteorite EETA 79001 and the other Martian meteorite DaG 476 receives a high (H) rating. These results probably reflect the high content of extractable nitrate and phosphate in these materials, maybe because there was no water in the cooling magma on Mars to leach out the soluble nutrients when these materials formed.

The carbonaceous chondrites Allende and Murchison also received a medium fertility rating, similar to a lava ash that is used by NASA as a lunar simulant similar to a sample collected by the Apollo astronauts. Such lava ashes are usually fertile soils. Interestingly, all of the planetary materials (non-Earth material) that we tested received higher ratings than a productive agricultural soil from New Zealand. This suggests that the asteroid and planetary materials are at least as fertile as productive terrestrial soils.

In summary, materials in planetary objects such as carbonaceous asteroids, cometary nuclei and interstellar dust particles are made mostly of silicate rocks, organics similar to soot and coal, and ice. These space materials are similar to materials on Earth, and their biological fertilities are also similar. These straightforward results have far-reaching implications: If Life can flourish on Earth, then Life can flourish throughout the universe.

Chapter 2.10 The Ecology of Solys: How Much Living Matter Can the Solar System Support?

The amount of biomass in any ecosystem, including planets and the galaxy itself, depends on:

- The amount of resource materials;
- The relative concentration of nutrients in the resource materials and in biomass;
- The rate at which this material is used up or wasted;
- Energy sources and living space that support the biomass;
- The lifetime of the habitat;
- In the case of human inputs, the effects of technology and its ethical basis.

To quantify the prospective amounts of life, we need units of measurement. The following analysis uses time-integrated biomass as a measure. There is of course more to life than biomass, but other aspects such as the complexity and quality of life are harder to quantify. As for the human prospects, we hope that conscious life and happiness will be proportional to the amount of living matter.

Nutrients and Potential Biomass

The amount of sustainable biomass is determined by the relation between the available nutrients in the resource materials and the nutrients required to construct biomass. This relationship can be expressed by $m_{x,biomass} = m_{resource} \, c_{x,resource} / c_{x,biomass}$, where $c_{x,,resource}$ (g/kg) is the concentration of element x in the resource material. The term $c_{x,resource}$ may relate to the concentrations of elements in the resource materials that can be extracted by water or by the roots of plants, or to materials that can be released by more advanced processing, up to using the total elemental content of the resource materials. Similarly, $c_{x,biomass}$ (g/kg) is the concentration of element x in a given type of biomass (eg., bacteria, plant or mammal), and $m_{x,biomass}$ (kg) is the amount of biomass that can constructed from an amount $m_{resource}$ (kg) of resource material based on element x. The element that allows constructing the smallest amount of biomass limits the overall biomass. The value of $m_{resource}$ using the carbonaceous asteroids and comets is known only as order-of-magnitude estimates. The calculated quantities of biomass in Table A4 in Appendix 1 are quoted to higher accuracies to illustrate the data that may be obtained when the masses and compositions of the resource materials will be known with enough precision.

Another factor of uncertainty is that $c_{x,,resource}$ will be modeled by measurements on the Murchison CM2 meteorite, but carbonaceous chondrite asteroids and comets may have different concentrations of these elements. For example, the concentrations of most extractable nutrients in the Allende CV3 meteorite are much lower than in Murchison, except for the important limiting nutrients N and P. Interestingly, these nutrients are also high in the two Mars meteorites but low in the lunar simulant lava ash. An encouraging sign for space materials is that the bioavailable concentrations of the nutrients in the meteorites are higher than in productive agricultural soil (Table 2.1)

The biomass that can be constructed from asteroids can be calculated based on measurements of $c_{x,resource}$ in representative meteorites (Mautner 2002a, b). The measured concentrations in some meteorites are listed in Table 2.1, and the concentrations of the same elements in several types of biomass are listed in Table 2.2. These data were used in equation (A1) in Appendix 2.1 to calculate the amounts of biomass (in grams) that can be constructed from the content of nutrient x in one kilogram of the asteroid/meteorite materials. The results, shown in Table 2.2, can be multiplied by the estimated 10^{22} kg carbonaceous asteroid materials to calculate the total biomass that can be constructed from the carbonaceous asteroids.

Estimated Populations

Table 2.2 shows that nitrogen is the limiting nutrient for constructing any types of biomass examined from either the extractable or total contents of the Murchison meteorite, followed by phosphate and potassium. For example, it is possible to construct only 0.34 grams of bacterial or 0.25 grams of human (mammalian) biomass from the extractable nitrate in 1 kg of the meteorite material. Table 2.2 also shows that the biomass can be substantially increased if the total elemental contents in the asteroids become available. For example, the bacterial biomass becomes 42 g/kg and the human biomass becomes 60 g/kg, increased over a factor of 100, compared with the extractable materials only. Interestingly, the total contents of each of the main biological elements C, H, O, N, P and K, as well as water, in Murchison would allow constructing comparable amounts of biomass. In other words, the relative concentrations of these elements in biomass are comparable to those in the carbonaceous chondrites. This may suggest that extractable materials in the carbonaceous asteroids/meteorites were produced by microorganisms, or conversely, that the composition of biological materials reflects their origins in carbonaceous asteroids or meteorites.

TABLE 2.2 Full wet biomass (g/kg) that can be constructed from each extractable element in carbonaceous asteroids or comets. *

	C	H	O	N	S	P	K
Bacteria	13	108	123	0.34	5736	0.7	12
Mammals	15	110	125	0.25	1900	0.5	181
Human Brain	14	112	122	0.32	4537	1.7	117
Average Biomass	16	112	119	0.48	4099	1.3	40

* For example, the amount of carbon (C) contained in one kilogram of carbonaceous chondrite asteroid materials can yield 13 grams of bacterial biomass.

Based on the extractable elements in the Murchison meteorite and on elements in biomass and formulas, we can calculate the total biomass in kilograms that can be constructed from the 10^{22} kg asteroids. This can be calculated by multiplying the numbers in Table 2.2 by 0.001×10^{22} kg. For the biomass constructed from 10^{24} kg cometary materials, multiply these data by 0.001×10^{24} kg. Further data and notes are in Table A4 in Appendix 1.

With the above data, we can calculate the human population that can be constructed from asteroid materials. According to Table 2.2, based on water-extractable nitrate in the 10^{22} kg carbonaceous chondrite asteroids as the limiting nutrient, we could construct 2.5×10^{18} kg human biomass. Assuming that an average individual weighs 50 kg, these resources would allow a population of 5×10^{16} humans.

However, each individual on Earth must be supported by a large amount of biomass and organic reserves. The estimated biomass of 10^{15} kg on Earth (Bowen, 1966) supports a biomass on the order of 3×10^{11} kg in the present population of 6×10^{9} humans, in other words, 2×10^{5} kg of biomass supports each human. This biomass includes trees, which will not necessarily be part of space-based ecosystems. Optimized ecosystems such as agricultural units or hydroponic cultures in space settlements may be more efficient, and a supporting biomass of 1,000 kg per person may be assumed. According to Table 2.2, the limiting extractable nitrate nutrient in 10^{22} kg carbonaceous asteroids allows 4.8×10^{18} kg of average biomass that would support a population on the order of 4.8×10^{15} humans.

Even with these conservative estimates, the bioavailable materials in the carbonaceous asteroids can support over one thousand trillion humans. The populations can be increased further by supplementing the limiting nutrients nitrate and phosphate from other asteroids or planets. Moreover, cometary nuclei in the Kuiper Belt contains about a hundred times more mass than the asteroids, i.e., on the order of 10^{24} kg. The Oort Cloud contains about a hundred times more mass yet, i.e., on the order of 10^{26} kg materials which may have similar composition to the carbonaceous asteroids. Using the bioavailable materials in these resources would allow on the order of 5×10^{22} kg biomass in the form of 10^{21} independent humans or as 5×10^{19} humans along with their supporting biomass. If all the elements in the asteroids and comets are liberated, these numbers can be increased further by a factor of 100. The asteroids would then yield 6×10^{20} kg biomass with 10^{19} independent humans or 6×10^{17} humans with supporting biomass. Using all the elemental contents in the comets would yield 6×10^{24} kg biomass in 10^{23} independent humans or 6×10^{21} humans and their supporting biota.

It is notable that this profusion of life can be based on materials basically similar to sand, soot and ice, using current levels of technology and accommodating humans with essentially current physiology. The large populations in many independent habitats can allow divergent biological and cultural evolution and adaptation, possibly extending to independent existence in space itself. These developments can prepare the further stages of expansion during the next phases of the Sun.

The Effects of Finite Time, and of Wastage

Most studies of population ecology concern the biotic potential, i.e., the maximum growth rate of a population and the final steady state population as defined by the carrying capacity of the ecosystem. Because the carrying capacity concerns steady-state populations, it does not include time and irreversible waste as variables.

However, the duration of an ecosystem may be finite if the environment becomes uninhabitable or if the resources are exhausted or lost by wastage. In this case we may be interested in the overall contribution of the ecosystem to the totality of life over its habitable time-span. This quantity may be measured by the total integrated biomass that will have existed in the ecosystem over its duration. Quantitatively, this is given by the biomass integrated over time, denoted here by the acronym $BIOTA_{int}$ (Biomass Integrated Over Times Available) and calculated using the equation $BIOTA_{int} = \int M_{biomass, t} \, dt$. The integral is over the whole time-span from the time of start (t_o) to the final time (t_f) of life in the ecosystem, where $M_{biomass, t}$ denotes the instantaneous amount of biomass at any time t. The integrated biomass may be expressed in units as kg-seconds, or more conveniently as kg-years. The total human life in the ecosystem may be expressed in human-years, similar to the way person-years are used when calculating labor. In the case of populations resulting from human activities, the growth-rate (biotic potential) and the desired maximum population (carrying capacity) may be defined by technical or ethical, rather than natural factors. However, these design factors can be affected by natural limits such as the amounts of resources and the expected duration of the ecosystem.

The habitable time-spans of life in the Solar System, the galaxy, and of the universe itself are defined by physical processes. One factor that limits the integrated biota is the waste that may be unavoidable for biological processes that cycle matter and dissipate energy. For example, space colonies or individuals may irreversibly lose gases such as oxygen, carbon dioxide, methane, water vapor and solid materials, and dissipate heat into space. The effect of waste on biosystems must be considered in calculating the *time-integrated biomass*. The equations for calculating the biomass are shown in Appendix 2.1. Assume that advanced technology will in a few centuries convert all the elemental contents in 10^{22} kg carbonaceous asteroids to 6×10^{20} kg of biomass including 3×10^{18} self-sufficient humans. Given the resources, this large population can be obtained from the current world population of six billion in 1012 years with an ordinary growth rate of 2% per year.

This advanced technology may also achieve a highly efficient system of waste minimization and recycling, where for example, only one part in ten thousand, i.e., 10^{-4} of the biomass is irreversibly lost each year. At this small rate of waste every kg of biomass yields 10^4 kg-years of time-integrated biomass. The time-integrated biomass based on the asteroid resources would then be $6 \times 10^{20} \times 10^4 = 6 \times 10^{24}$ kg-years. If all the biota were humans, the population will have lived for 3×10^{23} human-years compared with 6×10^{28} human-years without wastage. If comets can be also used as resources, these numbers may be multiplied by a factor of one hundred.

However, even at this low rate of wastage, the initial population of 6×10^{19} would decrease to the last human after only 455,386 years. The total biomass of 6×10^{20} kg would decrease to the last bacterium of 10^{-15} kg after 823,782 years, after which life would cease in the Solar System.

46

However, it is not necessary to construct and then rapidly waste all of the allowed biomass. Can the time-integrated biomass be increased if the resources are used gradually? For example, construction of the biomass from the same resources may be extended over five billion years so that a biomass of $6\times10^{20}/5\times10^9 = 1.2\times10^{11}$ kg is constructed per year. Nevertheless, the total amount of biomass constructed is still 6×10^{20} kg and each kg of biomass yields, with the same rate of waste as above, a time-integrated biomass of 10,000 kg-years. The total time-integrated biomass would still be, as before, 6×10^{24} kg-years. However, the slow construction strategy would allow life to exist in the current phase of the Solar System through its habitable life span.

In general, the time-integrated amount biomass is $M_{biomass}/k_{waste}$ which depends only on the total allowed biomass and the rate of waste but not on the rate of construction. On the other hand, when the biomass is constructed and wasted at compensating rates, the amount of steady-state biomass at any given time depends on the rate of construction. For the above example, the steady-state equation (A3) in Appendix 2.2 shows that if the asteroids are converted to biomass at the rate of 1.2×10^{11} kg y^{-1} and wasted at the rate of 10^{-4} $M_{biomass}$ kg y^{-1} then the steady-state biomass at any time would be 1.2×10^{15} kg. This biomass can include up to 1.2×10^{13} self-sufficient humans or 1.2×10^{12} humans each supported by a biota of 1,000 kg biomass. The latter population of over a trillion can permanently fill thousands of country-sized space colonies and planetary settlements. If humans are self-sufficient without supporting biomass, the population can consist of ten trillion space-adapted individuals. Furthermore, with cometary materials the population may be multiplied by ten thousand and the integrated $BIOTA_{int}$ will be 6×10^{28} kg-years.

With zero waste or with a rate of wastage less than $10^{-10}M_{biomass}$ per year, all or most of the biomass constructed in the next five billion years can survive the duration of the Sun. The total time-integrated biomass will be proportionally larger as k_{waste} decreases. From this point of view, planets where gravity completely traps the wastes for recycling, or perfectly sealed colonies, may be preferable to free individuals in space who could not avoid some wastage.

Energy

The maximum biomass sustained by solar power can be calculated using $M_{biomass} = P_{solar}$ x (efficiency)/(power use per unit biomass). Here P_{solar} is the total solar power output, 3.8×10^{26} Watts (which yields a flux of 1.35 kW/m^2 at 1 au). Plants can convert solar energy to chemical energy through photosynthesis at a theoretical efficiency of about 0.3 and an actual efficiency of about 0.05. The efficiency of current solar photocells is about 0.1 for converting solar power to electricity. In round figures, we can consider an efficiency of 0.1 in converting solar power for biological use both by photosynthesis and industrially.

The biological power needed by humans is on the order of 100 Watts per person (about 8,000 kJ/day) i.e., about 2 Watts per kg of total biomass or on the order of 10 Watts per metabolically active biomass, which requires 100 Watts of power supplied per kg of metabolically active biomass including the conversion efficiency. The power use in industrial societies is on the order of 10,000 Watts/person or 200 Watts/kg human biomass. Altogether, allowing for conversion efficiencies, the present estimates will use 100 Watts per kg biomass.

Therefore, the Sun can sustain with energy on the order of 4×10^{24} kg of biomass consisting of 8×10^{22} self-sufficient humans. Alternatively, if each human requires 1,000 kg supporting biomass, solar power can supply energy to a population on the order of 4×10^{21} humans. If the energy needs are modeled by industrialized nations, then the power requirements will be 10 kW per person and solar energy could support a population on the order of 4×10^{22} at industrialized standards. Note that these amounts of biomass and population use the entire power output of the Sun captured in a Dyson sphere. The biomass that solar power can support according to these estimates is comparable to or greater than the maximum available using asteroid and cometary materials, which is unlikely to be fully realized. Therefore energy is not a limiting factor of biomass and population in this period of the Solar System.

While discussing solar energy, we note that the intensity of solar light (energy flux per unit area) is an important biological variable. This flux varies inversely with the square of the distance from the Sun. Can photosynthetic algae and plants grow in the reduced solar light flux at the asteroid belt and in the outer Solar System?

As for algae, endolithic species can grow inside rocks and benthic species grow under water where only a small fraction of the solar radiation incident on the surface penetrates. As for plants, we tested the effects of reduced light intensity on asparagus tissue cultures. In growth chambers designed for tissue cultures, plants were grown at about one tenth of the natural solar irradiance, an intensity that corresponds approximately to solar radiation at 3 au, about the distance of the asteroid belt from the sun.

In addition, asparagus cultures were also grown on extracts of the Allende and Murchison carbonaceous chondrites and on the DaG 476 Martian meteorite at further reduced light flux, weaker about 80 times than the solar irradiance on Earth. The asparagus yields on the Allende extracts were reduced by a factor of 0.55 and on extracts of the DaG 476 Martian meteorite by a factor of 0.68, while those grown on the extracts of the Murchison meteorites increased by a factor of 1.25 compared with the cultures grown in full light intensity of the growth chambers. The light intensity in these experiments was comparable to the solar radiation at 9 au, about the distance of Saturn.

These results show that light intensity is a significant variable in astrobiology. However, solar light can support plant growth, if at somewhat reduced yields, even at the distance of Mars, the asteroid belt, Jupiter and Saturn. Of course, we know that many shade-tolerant plants can thrive in reduced light. Incorporating their light efficiency into crop plants can allow food production under a wide range of planetary lighting conditions.

Living Space

Assume that the largest population allowed by cometary resources, on the order of 10^{23} humans, is distributed about the Sun on a "Dyson Sphere" with a radius of 1 au i.e., 1.5×10^{11} m, at the distance of the Earth from the Sun. Assume that each individual will require the spacious living volume of a cube 100 meters on each size, about the size of a 40 story high-rise building. The required volume of 10^{29} m^3 can be provided by spreading out the population over the sphere at 1 au with an area of 2.8×10^{23} m^2 in a shell with a thickness of 354 km, about the length of a small country. The sphere may be of course centered closer or further from the Sun depending on the desired equilibrium temperature. Even much larger populations could easily be accommodated by spreading the population over thicker shells in the habitable zone. Metals in asteroids can easily provide the needed construction materials to house such large populations. Therefore, living space in the habitable zone also does not limit the population in the Solar System.

3

SEEDING THE UNIVERSE

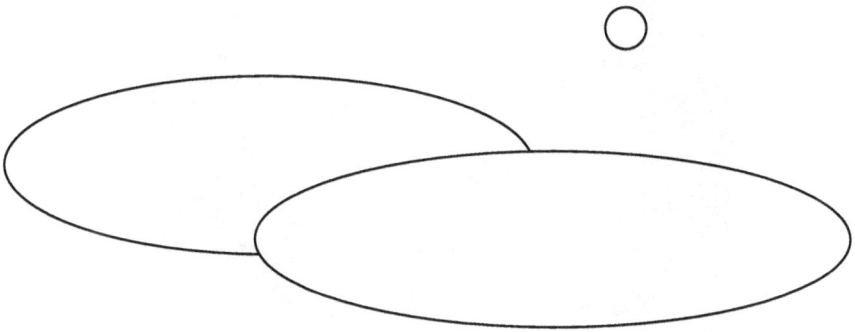

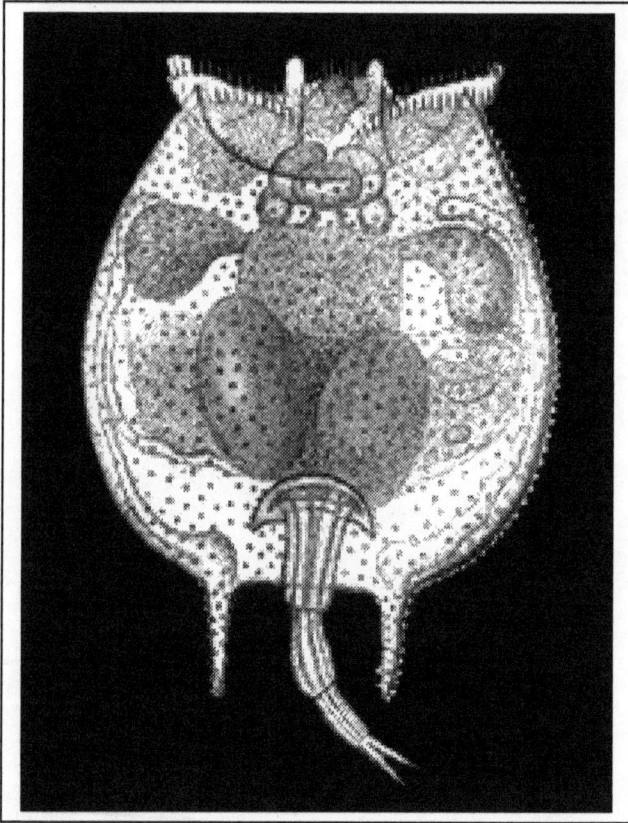

Representatives of our gene/protein life-form can soon start life in other solar systems. The pioneer organisms must be durable and versatile: bacteria, algae that produce oxygen and organics, and even multicellular organisms to jump-start higher evolution. This rotifer has hardy cysts for survival, and its body plan is suitable to start evolution leading to higher animals and maybe intelligent species.

Chapter 3.1 The Panbiotic Program: An Overview

*T*he objective of directed panspermia is to expand and maximize life in the universe.

Life on Earth is vulnerable and limited, but life in multiple worlds will be secure and virtually limitless. We have the means to secure life and disperse it into many worlds. Our civilization that can do that may last for ages, or it may decline in decades. We should act promptly.

Is there already other life in the galaxy? We may not know for a long time. Even a single cell is immensely complex, and life may be very hard to start. Our chances to interfere with other life are small, and we can further minimize this chance by targeting panspermia missions to young solar systems where advanced life could not have emerged yet. We cannot hold back from promoting life because of aliens who may or may not exist. Our duty is first to secure our own family of gene/protein life. If there is life elsewhere, we shall help to expand it; if there isn't, it is our duty to do so.

We can soon send missions to seed the nearby galaxy. We can launch solar sail missions to nearby solar systems where new planets are accreting. We can also send swarms of microbial capsules to star-forming clouds, where every mission can seed dozens of new solar systems. The materials needed for successful missions are available and manageable. The required technologies, precise astrometry for targeting and advanced interstellar propulsion, are advancing rapidly.

Biotechnology will help to design hardy and versatile organisms for panspermia missions. We can include algae that will produce their own organics, and even rotifers to jump-start advanced evolution. Our Solar System contains enough resources to seed every new star that will form in the galaxy during the five billion future years of the Sun.

Our missions can start multiple lines of evolution that will lead to diverse life-forms. Some of these future species may be intelligent and enjoy consciousness. Their descendants, and ours, will value and propagate Life further in the galaxy.

Chapter 3.2 Aims and Ethics of the Panbiotic Program

Aims and Ethics of the Panbiotic Program

As living beings, it is our purpose to propagate life throughout the universe. To pursue this purpose, humans similar to us will settle the Solar System. We hope that our descendants will continue to move on to other stars. However, interstellar human travel has major obstacles and its prospects are uncertain.

Nevertheless, we understand the importance of expanding Life beyond the Solar System. Where human beings cannot go, microorganisms can reach. They will represent us, as the essential patterns of our family of gene/protein life are contained in every cell.

Microorganisms are easy to transport, they can be frozen and desiccated, and revived after millions of years. They can survive in extreme environments of new planets and can evolve into higher life forms. Some of the new branches of life that we plant can lead to civilizations that will expand Life further in the universe.

The objectives of directed panspermia are to assure that the gene/protein form of life will continue, and to maximize life in the accessible universe.

Why Should We Secure Life?

Life on Earth may last for eons, but this planet will become inhospitable when the Sun ends in five billion years. Other catastrophic events such as major asteroid impacts may destroy all advanced life earlier. Our society has also developed the means to destroy itself, along with all higher life that could develop intelligence in the future. Indeed, living with our own means of self-destruction is the main challenge of the future. If any of these events happen, Life, if it is confined only to Earth, will vanish.

Yet there is room to secure Life for much longer. Life in the Solar System, and certainly in the galaxy, may last for many billions or trillions of years and maybe to eternity, depending on factors of cosmology that will not be known with certainty for eons. The first four billion years of life on Earth were just a short first instant of the potential future, and the number of organisms that have enjoyed life is only a minute fraction of life that may yet exist.

Life on Earth is vulnerable and limited; but Life in multiple worlds will be secure and virtually limitless.

A Genetic Space Ark: Why Now?

If we cannot be sure if Life already exists elsewhere in the galaxy, then we must make it sure that it will. How long shall we be able to do so? Only a few decades passed since humankind developed the technology to spread Life in space. In the same short time, humankind acquired the ability of self-destruction. Will our technology last for decades, centuries, or eons? We cannot foresee. We must act to promote Life while we are sure that we can do so.

It would be preferable for us if Life could colonize the galaxy in our human form. However, interstellar human travel has major physical, biological and psychological obstacles, and the required technology may not be available for millenniums, if ever. Even if human colonists leave for other stars, we may not know for millions of years if they succeeded. Human colonization is unlikely to secure Life in the foreseeable future.

We first proposed directed panspermia as a "genetic Noah's Ark" because of the buildup of nuclear weapons. We must recognize that the technology which can secure Life may also cause a global catastrophe. It is impossible to calculate the probability of such a catastrophe, but the panspermia project will be decided upon the *perceived* probability that life on Earth may be threatened. The survival time of the current civilization has been estimated by several authors as tens to thousands of years. A poll of students estimated the longevity of our civilization as 100-200 years (See 1979 panspermia paper, Appendix A3.3), and the astronomer Sir Martin Rees recently estimated only a 50% chance that our civilization will survive a hundred years. Given these estimates, it is essential that we secure life by seeding other habitats, and colonize space and other planets as fast as possible.

A further motivation for directed panspermia is that we can influence natural history on a cosmic scale. Beyond transforming the target ecosphere, the descendants of the pansperm program - if not our own descendants - may further spread life in the universe. Via panspermia we may thus ultimately contribute to turning life into a determinant force in the physical evolution of the universe.

Finally, the growing perception of the magnitude of the cosmos and the absence of evidence for extraterrestrial life so far, tend to induce a growing sense of our cosmic isolation. While the search for extraterrestrial life may lead to a passive solution, engineered panspermia will provide an active route of escape from the stark implications of cosmic loneliness.

We cannot predict how long our technological civilization will exist. We must secure and promote Life while we are sure that we can.

Does Life Already Exist Elsewhere in the Galaxy?

While millions of habitable worlds may exist in our galaxy, we don't know the probability that life may have originated elsewhere. Even the simplest self-replicating cell is extremely complex, and the probability that it could assemble spontaneously elsewhere in the galaxy may be miniscule. Life arose here within some one hundred million years after the earth became habitable. This is a short time astronomically, but in absolute terms one hundred million years are very long, allowing for trillions of steps of chemical reactions. It is therefore possible that only one in many trillions of possible reaction sequences can lead to self-replicating systems. This improbable sequence may not occur even on billions of other worlds. In any event, we cannot recreate how life actually arose and therefore we cannot calculate the probability that life will originate elsewhere.

In a few decades we may know if there are microorganisms on other planets, or on the asteroids and comets. However, it will be much harder to obtain evidence for life outside this Solar System. Any evidence in the next millenniums will be tentative at best as direct searches will require interstellar probes to explore many alien stars, each with transit times from tens of thousands to millions of years. Does Life exist elsewhere in the galaxy? We may not know for millions of years. Will our civilization last long enough to find out? We cannot predict.

We cannot recreate how life started and cannot compute how probable it is that life will start elsewhere. Our panspermia missions can assure however, that life will exist throughout the nearby galaxy.

Is Life Already Spreading in the Universe?

Large amounts of materials are being transported amongst the planets. For example, if the approximately 100 kg of Martian meteorites represent 0.1% of the infall during the last 10,000 years, then with an infall of 10 kg per year about 40 million tons of Martian meteorites would have fallen on Earth. Similarly large amounts may have been ejected from

Earth and traveled to other planets. Many of these meteorites are ejected relatively unheated and can carry viable microorganisms. There are speculations that some of the ejected meteorites that contain microorganisms could have been captured by comets and ejected in these cold storage containers into interstellar space.

Earlier in the Solar System, carbonaceous asteroids contained liquid water. These asteroids also contained organic molecules and nutrient salts that could have helped life to originate and would have supported microorganisms once they were present. Microorganisms can live on these materials, as was evidenced by cyanobacteria that colonized interplanetary dust particles in Greenland and by our experiments with bacteria and algae grown in meteorite extracts. As well, the early asteroids experienced many collisions that could have distributed microorganisms in the early Solar System. Some of these materials may be incorporated in comets, which would have brought them later onto planets.

Life could therefore spread effectively within a solar system like ours, once it is present there. Moreover, many meteorites, asteroids and billions of comets were ejected from the Solar System and they could have traveled to other habitable solar systems. However, because of the vastness of space, the probability of these objects landing on habitable extrasolar planets was estimated by scientists such as Fred Adams, Greg Laughlin and Jay Melosh to be very small. Therefore the probability of natural panspermia in the galaxy may be very small, and its pace very slow. Colonization by this process has been estimated by Adams and Laughlin to take longer than the habitable lifetime of the galaxy. In any case, it will be as hard to ascertain if terrestrial life is spreading in space as it is to find extraterrestrial life in the first place.

We cannot be sure if life exists elsewhere in the galaxy, and we cannot even be sure if we can find out. If life exists elsewhere, we can assist it to expand. If life does not exist elsewhere, we must assure that it will.

Does Intelligent Life Exist Elsewhere?

If the existence of life elsewhere is uncertain, the existence of intelligent life is even more so. First, we cannot calculate the chances that life will originate elsewhere. Secondly, prokaryotic cells (cells without a nucleus) evolved into eukaryotes (cells with a nucleus) only after billions of years, suggesting that this step is highly improbable and may not occur elsewhere. In this case, multicellular organisms that require nucleated eukaryotic cells could not evolve. Furthermore, the eukaryotes evolved into humans after six hundred million years, through many dead-end branches. Of the many millions of species and thousands of higher species, only one evolved technological intelligence. This suggests that the probability of the evolution of intelligent life may be very small.

The existence of extraterrestrial intelligence would be proven if we contact alien civilizations, or they contact us. However, sending and receiving interstellar signals requires a major commitment by us and by the putative aliens for many millenniums. The signals must be directed toward the recipients whose locations are unknown, must be in a form that unknown recipients will understand, and must arrive when the unknown recipients are listening. All this requires improbable coincidences.

If extraterrestrial civilizations do exist, some are likely to undertake galactic exploration and colonization. An advanced civilization can colonize the galaxy in a few million years. If such space-faring civilizations exist, we would have been visited already, but we have not. This is called the Fermi paradox.

All said, we cannot calculate with any certainty if intelligent life exists elsewhere. We are searching, but we cannot plan the future by waiting for the results. Certainly, we cannot endanger the future of Life by waiting to find out whether life already exists elsewhere.

While we hope for a contact, we cannot endanger our future by waiting for millenniums to search for extraterrestrial intelligence.

Will We Interfere with Other Life Forms?

What if our directed panspermia payloads fall on inhabited planets? First, we can minimize these chances by targeting star-forming zones or young solar systems where life still had little chance to arise, much less to evolve into higher forms. If there is primitive local biota, we may replace it by our messengers who have greater chance to evolve into higher life-forms. An exchange of genes between other gene/protein life-forms and our messengers will also help higher evolution. If our messenger organisms do fall on worlds with advanced life forms, then these locals will be able to defend themselves.

At worst, our messenger organisms may compete with other biota. Competition is essential to the advancement of life, and expansion in space will only extend it to a cosmic scale. While we will try not to interfere with other biota, our first obligation is to our own family of gene/protein organic life. We cannot abandon our existing family of life to an uncertain future on Earth and to certain death when our Sun ends, in order to protect other life-forms that may or may not exist.

As cognitive humans we wish to promote not only Life, but also intelligent life. We wish this not only for salient kinship, but also for the intelligent descendants of our program who may spread Life further in time and space in the galaxy. Therefore, we must send organisms that can survive the long journey, take hold in extreme environments and establish an evolving ecosystem with the potential to evolve into intelligent beings. Including primitive multicellular organisms in panspermia payloads can give billions of years of a *head start* to higher evolution, and may indeed be the only chance that this will occur in many habitats.

We cannot abandon our family of organic life to protect aliens who may or may not exist.

Should We Interfere With Nature?

If Nature has decreed that life on Earth shall end, should we try to save it? If Nature did not produce Life elsewhere, should we intervene?

Biology and intelligence are possible because of the laws of nature. Our intelligence and our ability to understand Nature and appreciate life were developed through natural evolution, not by our own design. Our ability to develop technology and promote life was also brought about by our natural development. The fact that space travel by human means is possible is also due to the laws of nature.

The impulses that we perceive as our free will, spring from human nature. Our strongest desire, is to continue life through natural procreation, is the core feature of life as developed by nature. Our desire for a cosmic future for ourselves and for all life results from our human impulses as formed by nature. We can fulfil this destiny with the means and to the extent that nature allows. It is the logic of Nature that the consequences of our actions are their own rewards: If we propagate life, our seed will survive; if not, we shall pass without consequence.

Nature brought forth life with an intrinsic purpose, self-propagation. We may never know, and can never prove by direct scientific observation, if the universe and life will exist forever. Nature may yet unfold so that we can implant life in other universes.

Nature brought forth our abilities to affect Nature. Nature brought us forth, set our path, gave us will and might. Nature will determine what future paths are possible, and the consequences of our future actions.

Willingly or not, we are a force of Life in Nature.

The Rationale for Directed Panspermia

Seeding the galaxy will influence natural history by human action at its best. The descendants of panspermia payloads may evolve and expand Life further, and help to elevate it into a dominant force in the universe.

Targeting young solar systems where advanced life could not have evolved will minimize interference with other life-forms. Our duty is first to our own family of gene/protein organic life. We cannot abandon our family of life for aliens who may or may not exist.

The magnitude of the cosmos and the lack of evidence for extraterrestrial life induce a sense of cosmic isolation. Directed panspermia will substitute this cosmic loneliness by the certainty of a cosmic community of living beings.

What do we know about life with scientific certainty at the present? We know that life exists on Earth but its future here is finite and uncertain. We understand that Life is unique and precious. We know that we can now plant our family of life in other solar systems where other intelligent beings may evolve who will further promote Life in the galaxy. We are a force of Nature that can expand Life in the universe. We are impelled to do so by the patterns of Life imprinted in our very being. If we advance Life, our human existence will acquire a cosmic purpose.

Directed panspermia will fulfil our human purpose to secure and expand the family of organic life in the universe.

Chapter 3.3 Directed Panspermia: A Technical Preview

This section previews some technical aspects of general interest. Specific technical subjects will be discussed in more detail in the following sections and also in the reprints of the panspermia papers in Appendix A3.3, A3.4 and A3.5.

Targets and Speed

In proposals of directed panspermia we considered two major types of targets. The first target may be planetary systems that are relatively close, say within 50 light-years from the Sun. Within this distance there are known solar systems with accretion rings or planets. Major efforts are underway to find solar systems with Earth-like habitable planets. If they are found we can check whether they harbor life by looking for telltale gases such as oxygen and methane in their atmospheres, and for the green color of chlorophyll on their surfaces. If we find planets that are habitable but uninhabited, they would be prime targets to seed with life by panspermia missions. However, accurately targeting small objects such as specific solar systems or planets is a significant challenge.

The second type of target is the star-forming zones in interstellar clouds. These are areas of relatively dense gas and dust which form new stars, their planets, asteroids and comets. These star-forming zones are large and can be reached with less precise targeting. In addition, each such zone typically forms 50 - 100 new solar systems simultaneously, which increases the chances that some will have habitable environments in which the panspermia payloads may be captured. Another major advantage is that life would not have been able to originate yet in these new environments and therefore we will not interfere with local life, and certainly not with advanced life.

As stated above, we wish to send our first missions relatively soon while we are sure to have the means and motivation. Launching these missions will be easy, probably within a century, at a time when there would be space colonies of idealist settlers imbued with life-oriented ethics. They will be able to readily launch from space solar sail missions that can achieve speeds of 10^{-4} c (One ten thousands the speed of light, i.e. about 3×10^4 meters/second).

Speed is a critical factor because we are aiming at moving stars and interstellar clouds, and their positions become increasingly uncertain with time. Also, long exposure to space radiation degrades the biological payload. At the realistic speed of 10^{-4} c, a solar system 10 ly (light-years) away will be reached in 100,000 years, and an interstellar cloud 100 ly away will be reached in 1 million years, well past the lifetimes of the senders and possibly of their civilization. Advanced technologies such as nuclear or ion drives and even antimatter drives may later reach speeds of 0.1 c, reducing the travel times by a factor of 1,000, but it is uncertain if these can ever be realized.

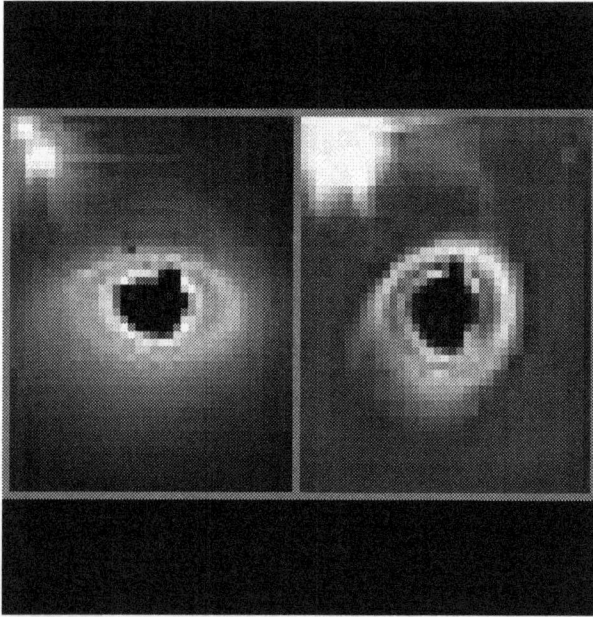

Dust disks around a young, 5-million-year-old star. Dust disks are expected to be the birthplace of planets. The star called HD 141569A, lies 320 light-years away in the constellation Libra. The image at left shows the disk slightly tilted when viewed from Earth by the NASA Hubble Space Telescope (left), and also as would be viewed from above (right). Panspermia capsules will be captured in accreting comets and asteroids in such rings and eventually delivered to planets.

Launching missions from space will be much easier and cheaper. Large infrastructures in space are expected to be constructed within the coming centuries. From space, solar sail missions can be launched by unfurling the sails outside space habitats to propel the sailships into interstellar space. A few dedicated individuals or small groups can launch this program.

In any case, directed panspermia missions can yield only moral benefits to the senders, and must therefore be inexpensive. Even so, the program must be motivated by selfless dedication to Life.

The Biological Payload

The biological payloads must survive the long trip in a deeply frozen, probably dehydrated state and must be able to be revived at habitable targets. It must include a variety of organisms, ensuring that some will survive in the diverse and extreme environments whose conditions cannot be predicted accurately. They must include algae that can develop into plants to support higher life. They must also include eukaryotes and small multicellular organisms that can evolve into animals and intelligent life forms. In this respect we are aiming to reproduce terrestrial evolution, while we recognize that on diverse targets, evolution may follow different paths.

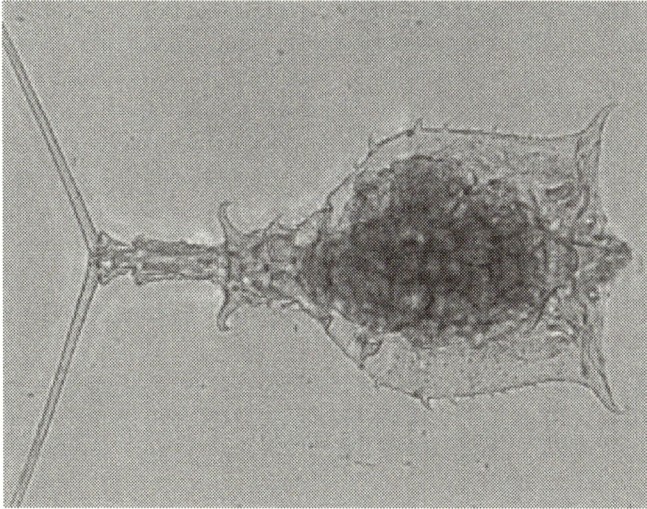

Multicellular animals in panspermia missions will increase the chances of advanced evolution. Rotifers are well suited as they have hardy cysts that may survive the journey, and a primitive animal body-plan that will allow flexible evolution.

Can solar nebulae serve as nurseries for microbial life?

Microbial capsule swarms are well suited for new planets and especially for interstellar clouds where they may disperse and seed many new solar systems simultaneously. While many will fall into stars or other inhospitable environments, some will arrive at colder zones where they can be captured by icy asteroids and comets and preserved until the objects or their fragments land on planets.

Some of the capsules may be captured in asteroids and exposed to water and nutrient solutions at the aqueous alteration stage of these objects. The larger asteroids of several kilometers in length contain liquid cores with abundant nutrients in which microorganisms can multiply. As the asteroids collide and fragment, they spew their liquid contents into space where it freezes. These chunks of ice will then be dispersed through the nebula and some will be captured in other asteroids where they can multiply further. Some of the microorganisms can be delivered to early planets at this stage. Other microbial populations will remain in the asteroids after they cool down; and will remain frozen and protected. Meteorites from these asteroids can then deliver the microorganisms to planets at a later time when life on them is at various stages of evolution. The first colonizers may actually create hospitable environments similar in the way cyanobacteria created the first oxygen atmosphere on Earth. Some of the multicellular organisms in the payloads that were preserved in deeply frozen asteroids and comets will be delivered to planets by meteorites later on, when the conditions are suitable for their survival. For this purpose we may include eukaryotes and multicellular organisms such as rotifer, tardigrade and brine shrimp cysts in the payload. The successful payloads can start new lines of complex evolution.

Some of the payloads will be captured in comets that are ejected into interstellar space. These comets can carry the frozen and protected payloads into deep space where they may be captured eventually into yet other new solar systems.

The following sections will address the technical aspects of the panspermia program in more detail. Further technical details are presented in the original papers that are reproduced in Appendix A3.3, A.3.4 and A3.5.

Chapter 3.4 A Closer Look at Directed Panspermia: Propulsion

As discussed above, we are looking primarily at methods to propel the panspermia missions that are relatively simple, based on known technology, feasible on a modest scale, and inexpensive. These criteria would seem to be difficult to satisfy for interstellar missions, but solar sailing can do so.

Solar sailing is based on radiation pressure. When photons from the Sun recoil from a reflecting surface, they impart forward momentum to the surface. Under this pressure, the object accelerates slowly but steadily, building up speed away from the Sun. The rate of acceleration is expressed by equation (1) in Appendix A3.3

Interestingly, If a sail is one atom thick, it could reach a velocity near the speed of light by the time it leaves the Solar System.

The critical factor for acceleration is the area of the sail on which the radiation impacts, which increases the force on the ship but also its mass. The ratio of these factors is the effective areal density m/A. For sails alone, thinner sails accelerate faster. If this factor is small, the radiation pressure can exceed the gravitational pull of the Sun and the spaceship will be accelerated away from the Sun. For example, calculations show that for a sailship with an effective areal density of 1.3×10^{-3} kilograms/m^2, the radiation pressure will exactly cancel the gravitational pull of the Sun at any distance from the Sun. Therefore, if the sail is unfurled in orbit about the Sun, the radiation pressure exactly cancels the gravity of the Sun. The ship will then cease orbiting the Sun and move off in a straight line with the same velocity that it had in its orbit. This is illustrated in the figure below, which also shows that if the effective thickness is smaller, the sail-ship will move away with an even greater speed. A velocity of 0.0001 times the velocity of light (10^{-4} c, or 30 kilometers per second, or 108,000 kilometers per hour) is easily obtained with a realistic sail thickness.

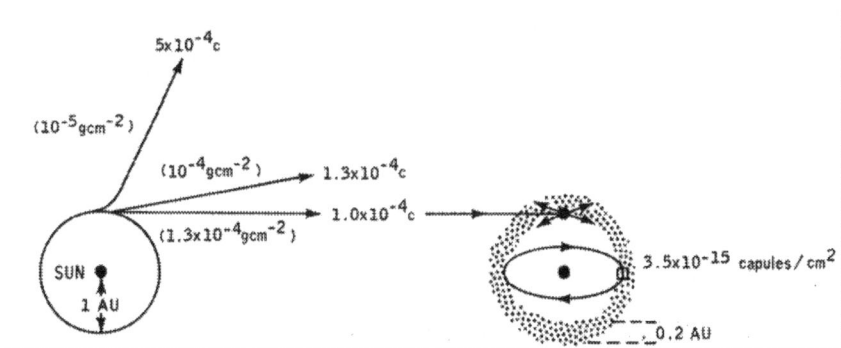

The launching of solar sail ships with effective thicknesses that will achieve final velocities as shown. The figure also shows the dispersion and capture of the microbial payload at the target solar system.

Nearby stars are typically 10 - 50 light-years away. A beam of light from Earth that travels at the speed of light (c) will reach these stars in 10 - 50 years. A ship that travels ten thousand times slower (at 0.0001 c) will reach the star after a time ten thousand times longer, that is, in 100,000 - 500,000 years.

During this voyage the vehicle will encounter interstellar gas that is very thin, on the order of one hydrogen molecule per cubic centimeter. Even so, an appreciable drag will be exerted during the long voyage. For example, for a trip of 10 light-years, the velocity will be reduced by 12 percent when the sail-ship reaches the target. The slowing effect will become larger on longer trips.

The drag could be reduced if the vehicle ejects the sail after the ship reaches interstellar space. However, an important advantage of the solar sail method is that the same sail will decelerate the ship when it is subject to the radiation pressure of the target star. Without the deceleration the spaceship would pass by the target star, but if it is decelerated it can be captured into orbit. The actual trajectory depends on the gravity and radiation pressure of the star, and on the properties of the ship. For this decelerating effect, the sail must be retained and protected during travel, so that the impinging gas does not erode it. It may be necessary to fold up the sail or use a venetian blind structure that will open up during the interstellar transit.

An important feature of the sail method is that the acceleration depends not on the mass of the sailship but on its effective areal density as discussed above. This applies no matter how small or large the object is. Therefore, appropriately designed solar sails can launch objects ranging in size from a microorganism of a few microns to milligram size packets of microorganisms, to capsules of several kilograms, or in theory an entire space colony if the sail is thin and large enough. We can take advantage of this property in the various panspermia strategies.

Chapter 3.5 Navigation

The optimal way to navigate panspermia missions would be to use in-course steering by robots. The required technology may become available in a few decades; these high-technology missions may require major resources.

For navigation, we must predict the positions of the targets at the time when the missions will arrive. To this effect, the positions and motion of the target objects must be known with sufficient accuracy.

The motion of stars may be decomposed into two directions: radial motion towards or away from the Sun, and perpendicular motion that changes the position of the star with respect to other stars as it appears in the sky. The latter is called the proper motion which appears as if the star moves typically across a very small arc of a certain angle during a given period of observation.

The proper motions of stars can be resolved by satellite-based measurements very finely, to 0.1 milli-arseconds per year. There are plans for a large array of 27 telescopes in space that will resolve star motions ten times better, to 0.01 milli-arcseconds per year. This motion is equal to a change in the position of the star by 7.7 parts per trillion of the full arc of the sky. The discussion below will assume this resolution.

Why is such an impressively high resolution necessary for aiming a mission? When an object moves in an arc as observed by us, the actual distance that it travels at its location is the product of the angle multiplied by the radius of the arc, which is the distance to the star. Although the angle is very small, the distance is very large, and their product is still an astronomically large number. Therefore a small uncertainty in the angle of the motion can translate to a large uncertainty in the calculated speed at which the star moves. Moreover, to calculate the uncertainty in the position of the star when the probe arrives, this uncertainty in the speed of the star must be multiplied by the long time of the travel to the star. The uncertainty in the position of the target at the time of arrival is given in Table 1 in Appendix A3.5.

We may assume that the error in targeting is oriented randomly and that the spaceships will arrive within a circle about the point that we target. For a simple approximation, we assume that the spaceships will be distributed randomly in this circle (More detailed models will assume a Gaussian distribution). We assume missions that arrive in the zone within a certain distance from the target, for example, within the area of a star-forming cloud in the habitable zone will be captured. The probability of capture can then be calculated from the ratio of the area of capture and the area of dispersion. We then obtain the relation in equation (2) of panspermia paper in Appendix A3.5, i.e., the probability for a vehicle travelling with a velocity v to arrive within a target area of radius r_{target}, whose proper motion is measured with a resolution of α_p is given by $P_{target} = (r_{target}^2 v^2)/(\alpha_p d^4)$.

In order to arrive at the area of a habitable zone about a Sun-like star, we must target an area with a size on the order of about 1 au, with a radius on the order of 10^{11} meters. In comparison, a dense fragment in an interstellar cloud is several light-years across, i.e., a radius ten thousand times larger, and its area is a hundred million times larger. In terms of the angle that the objects cover in the sky, a 1 au zone at 50 ly from us covers a minuscule angle of 0.000018 degrees in the sky, while the 6 ly star-forming zone in the Rho Opiuchus cloud 520 ly away distends an angle of 0.68 degrees as seen from Earth. Even the more specific area of a protostellar condensation in the dense cloud fragment in which a star is forming, is a million times larger than of a habitable zone. Therefore, we can target these areas with less demanding precision and yet with much more confidence. On the other hand, the swarm of microbial capsules will be dispersed into a much larger volume in the cloud and mixed with much more matter, of which only a smaller fraction will arrive eventually in a habitable environment. Calculating the overall probability of success must take all of these factors into account.

Chapter 3.6 Probability of Arriving at the Target Zone

The equations in Table 1 in Appendix A3.5 show that the probability of arriving in the target area increases strongly with the size of the target area. It also depends on factors that affect the uncertainty of the target's position when the vehicle arrives. These include the uncertainty in the motion of the target, the velocity of the vehicle and in particular the distance to the target.

For example, consider a mission to the habitable zone of a star. If the resolution in the proper motion of the target star is 0.01 milli-arcseconds per year and it is 10 light-years away, then the uncertainty in its actual speed is 4.88×10^{-10} light-years per year or 3.1×10^{-5} astronomical units per year (where 1 au is the Sun – Earth distance). If the spaceship travels at 0.0005 c, it will arrive in 20,000 years. In this time the uncertainty in the position of the star and its habitable zone will have accumulated to 0.6 au, that is, a little more than half of the distance from the Earth to the Sun. If the spaceship was aimed to disperse microorganisms in a wide ring about the star, the center of the ring will be positioned with this uncertainty in relation to the actual position of the star when the spaceship arrives. In this case the microbial capsules should be dispersed over an area with a radius of at least 0.6 au to make sure that at least some of the capsules arrive in the habitable zone. The need to cover a larger area because of this uncertainty requires sending more capsules and therefore the launching of larger or more vehicles.

Alternatively, we may aim at larger targets that are easier to reach. The Rho Ophiuchus interstellar cloud, or specific areas within the cloud, illustrates such potential targets.

In aiming for a star-forming cloud, we may target a dark cloud fragment or a specific protostellar condensation where stars are actually forming, in the dark areas shown in the Rho Ophiuchus cloud figure below. If we can achieve a high accuracy, we may target the actual disks of dust about new protostars where planets are accreting, at their earlier or later stages of accretion. Finally, we may even target planets themselves. We need to calculate the probabilities of success of each of these targets.

The parameters of the targets and the probability of success are summarized in Table 3.1. We considered solar sail missions with a speed of 5×10^{-4} c to the Rho Ophiuchus cloud at a distance of 520 light-years, and targets of various sizes in the cloud. The uncertainty in the position of the cloud and its components when the mission arrives will be 2.4×10^{14} meters, about one thousand times greater than the distance from the Earth to the Sun. Nevertheless, this uncertainty in position is only one percent of the size of the dense cloud fragment, and similar in size to the areas of protostellar condensation. Therefore, this uncertainty is acceptable in aiming for these objects. However, this uncertainty is ten times larger than the size of an accretion disk about an early star, and about a thousand times larger than the feed zone from which a planet collects material. Aiming for such objects will require the capacity to aim missions with high accuracies, where the uncertainty is comparable or smaller than the size of the targets.

Table 3.1 The probability that missions with the parameters described in the text will arrive at their target zones. Another probability for the eventual capture of the payload by a planet in the habitable zone about a star at the target was also calculated. Probabilities with a value grater than unity mean that that arrival or capture is virtually certain. Results for faster missions at 0.01 c give higher accuracies and better chances of arrival and capture. These are shown in Table 1 in Appendix A3.5 The paper also describes details of calculating the probabilities of success.

	Radius of Target area (au)	Uncertainty of target position (au)	Probability of arriving in the target zone	Probability of capture at a target planet	Biomass requirement (kg)
Nearby Stars					
Alpha PsA[c]	3.2 [d]	3	>1	1.1×10^{-5}	1.0×10^{-3}
Beta Pictoris[c]	8.7 [d]	16	0.3	2.5×10^{-6}	4.4×10^{-3}
Rho Ophiuchus Cloud[a]					
Dense fragment	200,000	80	>1	1×10^{-16}	1.1×10^{8}
Protostellar condensation	2,000	80	>1	1×10^{-13}	1.1×10^{5}
Early accreation disk	100	80	>1	3.9×10^{-14}	2.8×10^{5}
Planetary feed zone	3.5	80	0.002	4.9×10^{-11}	2.2×10^{2}

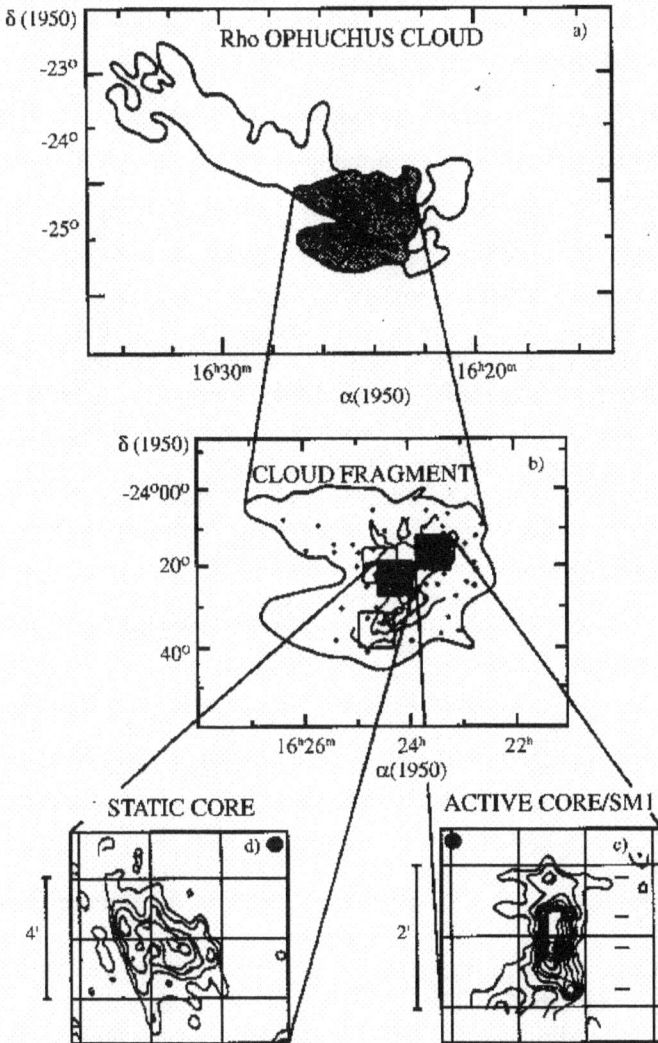

The Rho Ophiuchus cloud, a potential target for directed panspermia. The cloud contains areas of various densities. The dense fragment contains even denser protostellar condensations each of which will form a star, some orbited by planets. The cloud may give rise to a cluster of 100 new stars.

Chapter 3.7 Deceleration at the Target

After the vehicles or microbial capsules arrive at the target zone they must be captured. This requires that they slow down; otherwise they will bypass the target star or transit through the target cloud and continue on their interstellar journey without being aimed at any specific target. Their chances of landing on a suitable target in the vastness of space will then be minute.

In solar sail missions to other stars, the sails will be slowed down by radiation pressure in a symmetric but opposite from the way they accelerated away from the Sun. In an ideally symmetric case they would approach a Sun-like star at the same radial distance from the star as the orbit that they were launched from, and would be captured into a similar orbit. If the target star is different or if they approach it at a different radial distance, they may not be captured, or may be captured into different orbits. These trajectories require further research.

In missions to star-forming clouds we considered swarms of small capsules of 1 millimeter or less approaching the clouds. They may have traveled as separate capsules from the outset, or may have traveled in larger protecting vehicles and were released as separate capsules before entry to the cloud. Once they enter the cloud, the capsules will be slowed down by the gas and dust.

After entering the target interstellar cloud, the vehicles or capsules will first encounter a low-density envelope and then a dense cloud fragment before entering the densest star-forming core. In this core there are yet denser areas called protostellar condensations, each the site of the formation of an individual star. These condensed clouds will collapse further to form disks in which planets, asteroids and comets accrete about the new protostar. At each stage, the objects become more compact and dense.

Collisions with gas and dust particles in the cloud will decelerate the capsules or vehicles. The rate of deceleration depends strongly on their velocity (proportional to v^2) and on the density of the medium, and varies inversely with their size, i.e., larger and heavier objects will decelerate more slowly (see equations (3) and (4) in panspermia paper in Appendix A3.5).

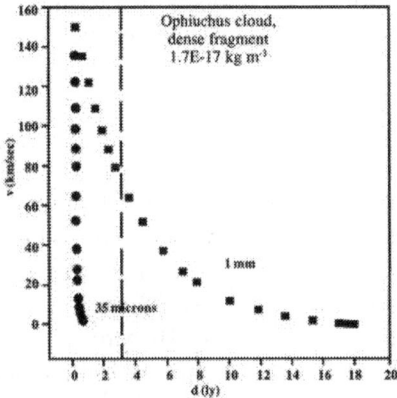

Ophiuchus cloud,
dense fragment
1.7E-17 kg m³

1 mm

35 microns

The deceleration of capsules with radii of 35 micrometers and 1 mm that enter a dense cloud fragment in the Rho Ophiuchus cloud. The graph shows the velocity as a function of penetration depth into the cloud, expressed as d (light-years). The vertical line shows the radius of the dense cloud fragment. The capsules can be considered stopped when their velocities decrease to 2 km/sec, similar to the velocities of dust grains in the protostellar condensation. The graph shows that the smaller 35 micron capsules are captured in the cloud fragment, while the larger 1 mm capsules traverse it and exit. Similar calculations for penetration into the more dense active core and protostellar condensation showed that the 35 micron capsules are captured in all of the target areas while the larger 1 mm capsules are captured only in the most dense protostellar condensation

The capsules in these calculations enter the cloud fragment with a velocity of 1.5×10^5 m/sec. Smaller capsules will stop after penetrating about one light-year into the cloud, while larger capsules will transit the cloud fragment and enter the active core. Both the small and larger capsules that enter the densest zones, called protostellar condensations, will be stopped there.

After sufficient penetration and deceleration, the velocities of the capsules will become similar to the velocities of the dust grains of the cloud itself, on the order 1,000 m/sec. At this point, the capsules become part of the cloud and have the same fate as its dust grains.

Chapter 3.8 Targets in Interstellar Clouds

If we can achieve enough accuracy, we can target zones of various sizes in the star-forming clouds. There are two opposing effects of aiming for smaller, more specific areas of star formation. On one hand chances are smaller for reaching smaller targets, but the capsules will then be scattered in smaller volumes and there will be more of chance for them to reach habitable planets.

We denote as P_{target} the probability for any given vehicle to reach the target zone, and as $P_{capture}$ the probability that once in the target zone, the payload will be captured into the habitable zone of a planet. The overall probability for capture in the target planet is then obtained from

$P_{planet} = P_{target} \times P_{capture}$. The combination of these factors will determine the biomass that we need to launch for a reasonable chance of success for missions to the various targets.

Dark Cloud Fragments

Because of the large sizes of clouds such as Rho Ophiuchus, virtually all of the vehicles or capsules will arrive within a targeted dark cloud fragment (the calculated value of P_{target} is actually greater than 1.) The dense zones in these fragments already have formed 78 stars, and we can assume that eventually it will form 100 stars using 10 percent of the total mass of the cloud. If the capsules are mixed evenly with the cloud, then thousands of capsules will be captured into each protostellar condensation that will form a solar system. From the rates of infall of interplanetary dust on Earth we estimate that a small fraction of the capsules, as little as 1 in 10^{13} of the capsules captured in each protostellar condensation, will be captured eventually as dust by a planet that forms in any of these solar systems during its first billion years. If the capsules were mixed with the dust in the cloud, and each star-forming protostellar condensation captures one thousandth of the mass of the cloud, then altogether 1×10^{-16} of the capsules will be captured by a habitable planet that may form in any of the new solar systems. If the number of habitable planets formed in the cloud is n_{planet}, then or a total of $n_{planet} \times 10^{-16}$ of the capsules will be captured by habitable planets in this new cluster of stars.

Individual Protostellar Condensations

These dense areas that will eventually form specific stars are still large enough so that each vehicle or capsule launched at them will be captured, giving $P_{capture} = 1$. As estimated in the preceding section, a fraction of 1×10^{-13} will be captured in a habitable planet, giving $P_{planet} = 1 \times 10^{-13}$.

The advantage of targeting individual protostellar condensations is that the capsules will be concentrated in a smaller zone, and mix with less material than if we aimed at the larger dark cloud fragment. On the other hand, there is a smaller chance of reaching these smaller targets and the mission to a particular protostellar condensation will not reach other stars that are forming in the cloud.

Early Accretion Disks

The young stars in Rho Ophiuchus are embedded in dust that forms a 100 au radius accretion disk. Because of their small size, the probability of reaching these targets is small, with P_{target} = 0.0039 for these objects. On the other hand, in the previous strategies the capsules were distributed in all the mass that forms the solar system, most of which will fall into the star where the capsules will be destroyed. Targeting an accretion disk avoids this major loss. Assuming that the majority of the dust is accreted into the original 1×10^{13} comets with a total mass of 1×10^{28} kg, and that 1×10^{17} kg cometary material is eventually captured by a planet, gives $P_{capture}$ = 1×10^{-4} and P_{planet} = 3.9×10^{-14}.

Accreted Planets

The most direct approach is to target planets in already accreted planetary systems. The best target planets are those that are at least half a billion years past their formation, when the conditions for survival become more favorable. We consider capture of vehicles or capsules that arrive within less than 3.5 au from the star, which yields P_{target} = 4.9×10^{-6}. From the statistics of the capture of meteorites and Zodiacal dust we estimate $P_{capture}$ = 1×10^{-5}, and therefore P_{planet} = 4.9×10^{-11}.

Chapter 3.9 Biomass Requirements

The amount of material that needs to be launched is calculated from the P_{planet} values that would deliver 100 capsules to each habitable planet. Assuming that each capsule contains enough microorganisms to seed a planet, the factor of 100 also corrects for other uncertainties in the mission. Of course, the smaller the probability for each capsule to reach a habitable planet, the more capsules we need to launch. In other words, the number of capsules that needs to be aimed at each target is inversely proportional to the probably of success. We calculate the biomass required for delivering 100 capsules of 1.1×10^{-10} kg each is then given by m = $1.1 \times 10^{-8}/P_{planet}$. The results are shown in Table 3.1.

For targeting the entire dense cloud fragment, a very massive program of 1×10^{8} kg per accreting star in the cloud is required. This is too much material to launch from Earth at current costs but could be achieved using asteroid resources. Producing this amount of payload requires the organic and water contents of about 10^{10} kg of carbonaceous chondrite materials, about a cube of 200 m on each side mined from an asteroid.

This material could be processed into microbial biomass and some protective coating in an asteroid base and launched from this low gravity object.

If we can target individual protostellar condensations or accretion disks, the requirements are a realistic 100,000 kg, requiring materials from as little as a cube of 20 m mined from an asteroid. Launching this mass could be also accomplished by Earth-based missions, especially if the launch costs are reduced.

Finally, if already accreted planetary systems can be identified and targeted in the cloud, or nearer in space, the mass requirements are reduced to the order of 1 to 100 kg, which can be met easily with current technology, and may be achieved at costs of less than a million dollars. Such missions would then be affordable to small groups or even individuals. Such low cost programs may be realized in this century.

Chapter 3.10 Swarm Missions to Nearby Stars

Microbial capsule swarms can be also sent to planetary systems that are closer than star-forming clouds. For example, the star AU Microscopium, a new red dwarf star only 12 million years old was recently found at a distance of 33 ly, who has a dust ring with a gap. This suggests the formation of a planet that is sweeping up the dust. Another nearby star is alpha PsA (Fomalhout), at a distance of 22.6 light years from us. Calculating the probability of arriving at the targets, missions to Fomalhaut gave P_{target} of 1.2, and missions to Beta Pictoris of 0.25, for capturing missions aimed at these targets into orbits in the habitable zone. For $P_{capture}$ we use $1x10^{-5}$, although this may be different in different solar systems. With this assumption, $P_{planet} = 1x10^{-5}$ and $2.5x10^{-6}$, respectively for the two targets. These stars are in the local low-density interstellar medium, and solar sail methods can be used to launch 30 micron radius capsules each weighing $1x10^{-10}$ kg, attached to miniature 1.8 mm radius solar sails. These sails may be, for example, envelopes of thin reflective film that enclose the payload.

Another possibility is to fold up the sail after launch, especially about small microbial packet payloads, so that they will form a protective envelope during interstellar travel. Moreover, the film may be made of biodegradable polymer that may serve as the first nutrients after the microbial packets land on habitable targets. The microbial packet surrounded by a protective nutrient coating will constitute a panspermia egg. These sails will have to be made of "smart materials" that respond automatically to various environments. They can fold up in the interstellar cold, unfold again to slow down the packet when approaching the target star, and absorb water on the target planets to serve as nutrients. Robots could help such missions but it would be hard to be sure that they can survive for tens of thousands of years.

The sails and capsules for the microbial swarm missions may be mass-produced using microencapsulation. As few as 1×10^7 to 10^8 capsules with a total payload of one to ten grams could then deliver 100 capsules to a planet. Remarkably, even with current technology, a panspermia swarm with a reasonable probability of success can then be launched to these stars, nominally, at launch costs of $10 per planet. With such costs the missions can be easily scaled up to kilogram quantities to increase the probability of success or to allow for less accurate, easier methods of launch and navigation, still at a nominal launch cost $10,000. These launches may be added as riders to other space programs, rather than requiring more expensive dedicated missions.

The technology to produce the microbial capsules and to launch them accurately requires research and development using currently available technology. The costs are affordable and panspermia missions to nearby planetary systems can be launched in this century.

Chapter 3.11 Designing the Biological Payload

The first objective of the panspermia program is to expand our family of gene/protein life. All cellular organisms contain the essential structures of cell/protein life, and therefore any organisms sent to the new habitats will serve this purpose.

However, as conscious beings we hold higher, self-aware life to have special value. Moreover, as all species, we desire to perpetuate our likeness. This serves the interests of life at large, as intelligent, technological beings are needed to expand life in space deliberately.

The biological payload should be able to:

(1) Survive the launch, the long transit in space and the capture.
(2) Grow on nutrients and conditions likely to be encountered under diverse planetary conditions.
(3) Initiate evolution toward higher life forms.

Scientific advances in several areas in recent years have contributed toward these goals. These advances include the discovery of extremophile microorganisms that can live under ever more diverse and extreme conditions; the observation that microorganisms can survive on extraterrestrial materials; and the advent of guided microbial evolution.

The diverse but potentially survivable conditions at the target planets may range from icy planets at 100 - 280 K and space vacuums, to the upper atmospheres of gas planets, to Martian or terrestrial environments, including hydrothermal conditions of 370 - 648 K (the critical point of water) and pressures of several hundred bars. It is encouraging that in the last decade microorganisms have been found that can survive in some of these extreme environments. Cyanobacteria and lichens can grow in cold dry Antarctic valleys that resemble icy comets and planets. Endolithic microorganisms that grow under layers of minerals in rocks at low light levels similar to those in the outer solar system and about dim stars.

Most new environments are likely to be anaerobic, and may be at elevated temperatures of 60 C to over 100 C.; they may contain carbon dioxide, hydrogen, sulfur and sulfides. Such environments include volcanic pools and hydrothermal ocean vents. In particular, such environments are found in asteroids under aqueous alteration and in the pores of meteorites that have fallen on early aqueous planets. Remarkably, archaebacteria that lie near the root of the phylogenetic tree are anaerobic heterotrophic thermophiles that are suited to just such conditions. Their features may in fact reflect conditions in the early Solar System from where they could have originated. Such conditions are also likely to be found in other early solar systems.

Other microorganisms can survive under high salinity, or in highly acidic conditions. Most remarkable is *Deionococcus radidurans* that developed a genetic mechanism to cope with extremely high doses of radiation. Cyanobacteria, which formed the first chert fossils on Earth and created the oxygen-rich atmosphere in which higher evolution occurred on Earth, are also useful early colonizers. Some still retain the ability for anaerobic metabolism, presumably from their early evolution. With these features they can be useful colonizers of other planets.

In order to initiate higher evolution on the target planets promptly, it would be desirable to include eukaryotes and even multicellular organisms in the payloads. Judging from terrestrial history, the progress to eukaryotes is a major evolutionary bottleneck, apparently of such low probability that it may not happen at all in most other ecosystems. A panspermia payload that includes eukaryotic and multicellular organisms can give a headstart of billions of years to higher evolution; in fact, it may be needed to ensure that higher evolution will occur at all. It would be also desirable to use organisms with a high rate of mutation and a mixture of organisms that will exert mutual evolutionary pressures on each other.

The inclusion of simple multicellular eukaryotes is crucial, as this may help bypass major evolutionary bottleneck. The evolution of single cell to multi-cell organisms required billions of years on Earth, but then led rapidly to higher life forms. Such a low probability event may not occur at all in other evolving ecosystems.

Some of the eukaryotes could be algae, which can give rise to higher plants. It may even be possible to include small seeds of hardy plants. As for animals, amoebae are the simplest organisms to start higher animal life. However, there are other microscopic animals that produce hardy eggs or cysts that may survive long periods of very low interstellar temperatures, vacuum and dehydration. In fact, these interstellar conditions may help to preserve the biological materials.

Rotifers may be particularly suitable candidates. These small animals consist of a few dozen cells that contain differentiated internal and external layers, with the internal organs fulfilling digestive and reproductive roles, similar to higher life forms. They contain the basic body plan of all higher animals, but their anatomy is still open to diverse body plans. Some of the species that diverge from them may evolve in directions that could lead to higher intelligence. Other microscopic animals that developed survivable cysts such as tardigrades and brine shrimp have already evolved as insects and invertebrates and are less open-ended for higher evolution. However, they can be also included to induce a diverse ecosystem.

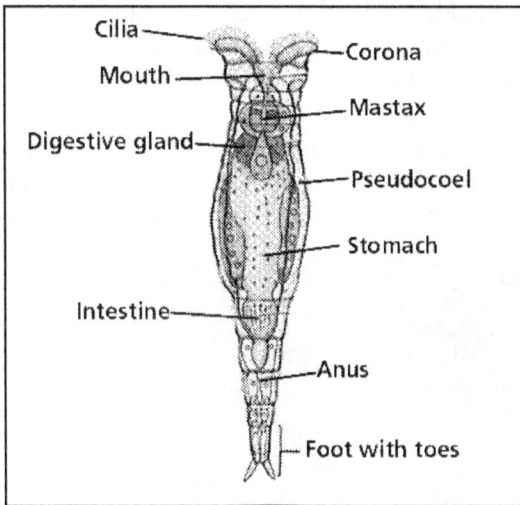

There is enough diversity in biology to include payloads of organisms with diverse biological capabilities. It would be even better if each organism in itself could be as hardy and versatile as possible. Such organisms can be developed by guided evolution. For example, guided evolution is achieved by amplifying bacterial DNA using "sloppy PCR" methods. Mutant DNA is introduced into microorganisms that grow under conditions that are altered gradually. Also, mixed cultures of organisms can be grown under conditions that facilitate horizontal gene transfer. This will help organisms resistant to heat, cold, salt, dryness and radiation to acquire these traits from each other. The organisms can be developed in planetary microcosms starting from terrestrial conditions and changing gradually to resemble diverse asteroid and planetary conditions.

The physical design of the microbial payload may benefit from what nature invented in the seed and the egg. Each packet of microorganisms can be encapsulated in a protective but water-permeable polymer shell containing a layer of nutrients that will help revive the freeze-dried payload upon rehydration after capture at the target planets. The medium provided as the first nutrients can mix gradually with the local nutrients and help the microorganisms to adapt gradually to local nutrients.

For interstellar transit, the microbial payload may be freeze-dried, as is the current practice for preserving microbial cultures. For UV survival, the capsules must be shielded at least with UV resistant films. It may also be desirable to include a nutrient medium in the capsule, and to enclose it in a selective membrane that will allow the supplied nutrient to slowly absorb planetary nutrients, so that the microorganisms can gradually adjust to the planetary chemistry (pH, redox potential, toxic components, and specific local nutrients). For aerobic eukaryotes, it may be desirable to enclose them in separate capsules with shells that will

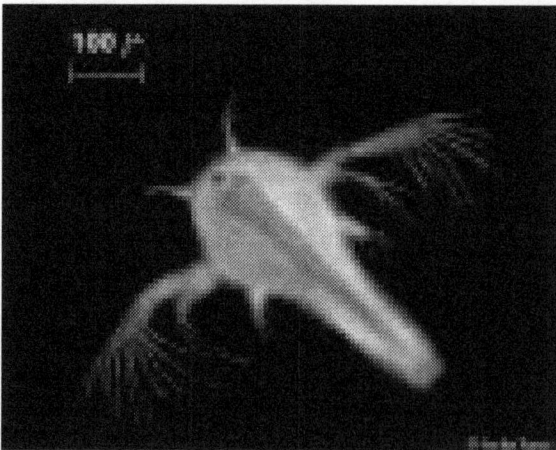

Brine shrimp. The hardy eggs of rotifers, tradigrades ("water bears") and brine shrimp may be included in panspermia payloads. We found that brine shrimp eggs can hatch in extracts of carbonaceous asteroids/meteorites.

dissolve only in oxygen-containing environments. This will preserve the aerobic eukaryotes until photosynthetic organisms in the panspermia payload create a suitable oxygen-containing atmosphere.

Chapter 3.12 The Fate of Microorganisms in The Target Environments

The first task of the biological payload is to survive the interstellar trip that may last from tens of thousands to millions of years. Can microorganisms survive that long? Some of the first experiments by Peter Weber and J. Mayo Greenberg have shown that the survival rates of microorganisms under UV light increased at low temperatures of 10 K, similar to interstellar conditions. Recent findings suggest that microorganisms can indeed survive for long periods of time. Ancient microorganisms that were trapped for thousands of years in amber have been revived. Most remarkably, Dr. Russell Wreeland succeeded in reviving the microorganism, *Bacillus permians* that was trapped in the liquid of a salt crystal for 250 million years. "It was completely protected" said Dr. Wreeland in a newspaper interview. "It was able to shut itself down into a protective spore and once it was encased in this particular type of rock it found itself in the most stable environment that you could imagine. If an organism were encased in a crystal and blown off a planet somewhere, or blown off this one....it has a reasonable probability of surviving long enough to travel not just from planet to planet but from solar system to solar system". We can even improve these chances by designing the best protection system; by including in such salt packets microorganisms that suit diverse conditions, and by including nutrients that will help the microorganisms establish themselves once they arrive in habitable environments.

As discussed above, some of the microbial payload capsules will be captured on early planets and colonize them directly. Other capsules will be captured in asteroids and comets and preserved in cold, dark conditions shielded from radiation. These capsules will be released from the asteroids and comets in meteorite debris or dust later and captured by planets after the first colonists rendered them habitable.

Several authors, especially Hoyle and Wikramasinghe, theorized about the possible roles of comets in dispersing life in the universe. Our plans take advantage of these ideas in an active manner. In fact, capture of the capsules in comets and later delivery to planets seem as the most likely mechanism for success.

It is widely recognized that comets may have delivered most of the water and much of the complex organics that formed life to the early Earth. The pores in a carbonaceous meteorite that fell on Earth could have, and still can today, fill with water and create in their pores a rich solution of organics and inorganic nutrients. These environments are suitable for prebioic synthesis. Meteorite pores are suitable in particular as the solution becomes trapped and allows many steps of the organic synthesis leading to complex large molecules. Microorganisms may be introduced into similar asteroid/meteorite solutions through directed panspermia missions.

Our research directly addressed the fate of microorganisms in solutions of meteorite materials, and on the wetted meteorites themselves (see astroecology papers in Appendix A3.1 and A3.2,). In the first experiments we found that organics in meteorites such as the Murchison carbonaceous chondrite, and presumably in comets and interplanetary dust particles, are released efficiently at about 10% at 20 C and up to 50% at 350 C under simulated early planetary conditions. The released organics were utilized efficiently by common environmental microorganisms. For example, the meteorite powder was suspended in water which was then plated on agar plates. Microbial colonies grew on the plates and were identified tentatively as *Flavobacterium oryzihabitans*. When the solution was incubated at 25 K for 6 days, it gave a population of 1×10^8 CFU/ml (colony forming units/ml), a thousand times higher than populations that grew in deionized water. These microorganisms utilized about 3 percent of the released meteorite organics. The humic acid utilizing bacterium *Pseudomonas maltophilia* gave populations of 1×10^6 CFU/ml in the meteorite solutions, about five times more than in deionized water. Another microorganism that was found in these cultures and gave especially good responses to the meteorite nutrients was the oligotroph *Nocardia asteroides*, a microorganism that grows well on small amounts of nutrients.

In a further step to simulate the ecology of planetary soils, we inoculated the meteorite extracts with algae and with an extract from a natural wetland. The cultures developed a mixture of microorganisms and some were sustained for over a year, which was the length of time the cultures were followed. These results are described in the astroecology papers in Appendix A3.1 and A3.2.

Much of the point of these experiments was to observe if potential panspermia payloads could establish a viable microbial ecosystem on carbonaceous chondrite materials. The results, together with the earlier tholin findings suggest that a proper microbial payload can be developed to survive on such materials.

It may be possible to use the solar sail that was used to launch the capsule to provide effective shielding as well as being an extra nutrient source. The sail will constitute about 90% of the total mass of the small vehicles and could contain proteins, carbohydrates, and other biodegradable organics. It may be designed to fold over the microbial packets after propelling them from the solar system to provide shielding during transit and capture, and eventually to provide nutrient materials on the host planet.

The first colonizers may be heterotrophs, microorganisms that use the organics already found in the environments. The phylogenetic tree points to such organisms as the first life on Earth. Later, after the local organics are used up, it will be necessary for life to create its own organics. Recent genetic evidence suggests that the photosynthetic apparatus that creates organic biomaterials in plants evolved from cell components of autotrophs that at first had other functions. This may also happen with our payload, but to make sure that self-sufficient organisms will be present, we shall include cyanobacteria and algae that are already equipped with photosynthetic capacity.

Chapter 3.13 Seeding the Galaxy

Although aimed at specific planetary systems and star-forming clouds, the microbial payloads may carry life further into space and time. First, much of the microbial swarm will miss or pass through the target. Secondly, of the initial 10^{13} comets that capture capsules in an accreting system, up to 99% may be ejected into interstellar space as it seems to have happened in our early Solar System. These comets will carry along their microbial contents. These embedded capsules, shielded from radiation and preserved at 3 K, may survive in the comets in interstellar space for millions, even billions of years, until some are captured in other accreting planetary systems in the galaxy. Of the 10^{11} comets remaining in the accreting system at the original targets, most will remain in the cold Oort cloud and be preserved in deep freeze at temperatures below 10 degrees Kelvin. Many of these comets will also be ejected eventually into interstellar space. Therefore the majority of the launched biomass will carry the microbial payload further into the galaxy. The spread of microbial life by comets is similar to the proposals of Hoyle and Wickramasinghe but with a directed origin.

Future programs may aim to seed the entire galaxy intentionally. We considered vehicles or capsules launched randomly into the galactic plane with speed of v = 0.0001 c or at 0.01 c if launched by advanced methods. The microbial packets will traverse the galaxy whose radius is 70,000 light-years, in the order of a billion or ten million years, respectively. The packets are gravitationally bound to the galaxy and will eventually take random paths. At these speeds, millimeter size capsules will transit through all thin regions and will be captured only in star-forming protostellar condensations or denser accretion zones of interstellar clouds. The mass ratios of total to captured material above showed that 10^{-13} of the captured biomass in these areas will be delivered to planets. With 100 capsules of 10^{-10} kg, a biomass of 10^{-8} kg is required to seed a planet in a new solar system, requiring us to launch 10^{5} kg of microbial capsules to seed each new planetary system. Given that about one star per year is formed in the galaxy, we need to launch a biomass of 10^{5} kg per year for the next five billion years to seed all new stars that will form in the galaxy during the lifetime of the Solar System. For example, the biomass can be dispersed in pulses of 10^{12} kg to seed the population of star-forming clouds as this population of stars is renewed every ten million years. The total required biomass is 5×10^{14} kg, compared for example with the 10^{20} kg of organic carbon contained in the 10^{22} kg of carbonaceous asteroids. Using all of this resource the launched biomass can be increased by a factor of a million above the estimated need to seed star-forming clouds, to account for losses of capsules in these missions.

As a more conservative estimate, assume a 5 au capture zone about each star with a volume of 2×10^{36} m^3, which gives a total capture volume of 2×10^{47} m^3 about 10^{11} stars. With a capture probability of 10^{-5} for delivering a capsule captured in this volume to the planet, if we wish to deliver 100 captured capsules of 10^{-10} kg each we need to place one gram of capsules about each star. This corresponds to a density of 5×10^{-4} kg biomass m^{-3} in these circumstellar volumes. Assuming that this is achieved by establishing a similar density of biomass through the entire 5×10^{61} m^3 volume of the galaxy, the total biomass needed in the galaxy is 2.5×10^{22} kg. We assume that this density needs to be renewed each billion years for the five billion year lifetime of the Solar System to seed every new planetary system during the first billion years after its formation. This gives a material requirement of about 10^{23} kg, about 10% of the 1% carbon content in 10^{26} kg of the total mass of comets.

The material requirements can be reduced by many orders of magnitude if the missions are directed to star-forming regions rather than distributing the biomass randomly through the galaxy. The microbial population may be subject to substantial losses, but its concentration in the accretion zones may be enhanced by gravitational attraction to these dense zones. The fate of biological objects traversing the galaxy requires detailed analysis.

It may be possible to grow the necessary large amounts of microorganisms directly in carbonaceous asteroids or comets. Carbonaceous C1 meteorites, and presumably asteroids, contain water, carbon and other essential biological elements in about the required biological ratios. The biological elements can be extracted and the residual inorganic components may be used as shielding materials for the microbial capsules.

If we can reach and process cometary nuclei, we can fragment the loose icy cometary matrix into one kg spheres, and enclose each sphere in membranes. Warming and melting such a unit from 10 to 300 K requires 5.1×10^9 Joules of energy. This energy can be provided by the solar energy flux of 325 W m^{-2} at 2 au that is incident on the 3.1 m^2 cross-section of a 1 m radius object during a two-month period when it approaches a sun. Our microbial experiments showed that in 6 - 8 days after inoculation this organic solution will yield microbial densities of $>10^8$ CFU/ml which can survive for several months (See the astroecology papers A3.1 and A3.2 in the Appendix). Subsequently, the microbial solution can be converted to 1 mm "hailstones". These microbial ice capsules can be accelerated out of the solar system by first accelerating the comets sunward into parabolic orbits. In this manner we would use up and disperse the material through the Oort cloud at the rate of 20 comets yr^{-1} during five billion years. This rate is comparable to the natural rate of 3 new comets/yr plus up to 10^9 new comets per year during cometary showers. Therefore, the Oort Cloud comets can be converted into microbial panspermia capsules at the required rate by processing every new comet that arrives naturally from the Oort cloud.

An interesting experiment would be to inoculate the zone under the crust of an inbound comet, or to embed samples of the cometary material in containers and allow them to melt when the comet approaches the Sun during its perihelion phase. Embedded sensors could monitor microbial growth during this period and during further passages about the Sun. Laboratory microbiology experiments with returned cometary materials would also be of interest.

The above considerations suggest that a single technological civilization in a single solar system can seed the galaxy. By extrapolation, the materials in ten billion solar systems in the galaxy would be sufficient in theory to seed with life all the ten billion galaxies in the universe.

Of course using comets in this manner is a prospect for the long-term future. However, missions to land on comets are being planned presently, and biological experiments to test their biological potentials can be implemented in the near future. Comets bound for interstellar trajectories could be seeded with microorganisms in a few centuries.

Chapter 3.14 Panspermia Programs by the Current Civilization

Parts of this book present ethical arguments to expand Life. The present section, based on previous published papers starting in 1979, outlined a panspermia program that appears to be the best way to expand Life beyond the Solar System in the foreseeable future.

This program is both feasible and prudent. We hope that the human future will be long and glorious. However, we realize that it is also uncertain. A panspermia program can assure that we make an impact that will last as long as the universe is habitable, through unimaginably long trillions of years into the future. We can then be assured that our family of organic gene/protein life will inhabit the universe. From the vantage point of Life, this is the most important deed that living beings can accomplish.

These prospects are within reach. The scientific basis exists, and the needed technologies of propulsion, astrometry, the search for extrasolar planets, extremophile microbiology and genetic engineering, are all progressing rapidly. An infrastructure in space is also building up, and from there, panspermia missions can be launched readily even by small dedicated groups. Seeding the galaxy with life will be an obvious moral duty to generations born in space.

Chapter 3.15 Advanced Missions

Our discussions have emphasized current technology that could allow panspermia missions within a few decades with affordable costs. Making the program dependent on future technologies can put them off indefinately and make their realization uncertain. Nevertheless, it is worthwhile to consider advances that may decrease the costs and increase the probabilities of success.

Solar sailing propulsion could be enhanced by propelling the sailships using lasers or microwaves. Other advanced methods include nuclear propulsion and ion drives which have been proposed for a long time. They may achieve higher speeds than solar sails, and as we saw previously the probability of targeting success increases strongly with speed and with the square of the velocity. Fast transit also decreases exposure to interstellar radiation.

Another concept that is receiving serious thought is the space elevator, a cable that moves cargo to a satellite in geosynchronous orbit above the equator. The cable is 47,000 kilometers long, made possibly from ultra strong carbon nanotubes. The space elevator could reduce the costs of launch from the order of $10,000 to one dollar per kilogram. Once elevated to space, launching microbial solar sail capsules will be cheap and easy.

Most of the described programs above, even those targeted at protostellar condensations and requiring 100,000 kg biomass, could easily be financed even by an individual. The probable success of other missions that need less biomass could still be enhanced by increasing the launched biomass, increasing the probability of success hundreds or thousands of times. Cheap launchings can also make missions to many more targets possible.

A new idea proposed by Dr Robert Winglee uses a solar wind of charged particles by creating a magnetic shield about the spacecraft similar to the field that surrounds the Earth. This method could propel a 100 kg spaceship to 10^{-4} c in a day, or to 0.01 c using three kg of helium to sustain the plasma for three months.

An even more exotic technology is to use antimatter - matter recombinations that could propel vehicles to near the speed of light. The energy needed to launch a one milligram capsule at 0.01 c is 4.5×10^6 Joules which can be provided by converting 5×10^{-11} kg of antiparticles to energy. Launching smaller, microgram capsules at 0.01 c requires the production of 5×10^{-14} kg of antiparticles, which brings even this exotic method within the capabilities of current technology.

In terms of navigation and target selection, a space telescope could be placed at the stable Lagrange 1 point. Together with Earth-based telescopes and maybe telescopes on the Moon, the resolution of astrometry, and therefore the accuracy of targeting, could be increased by a factor of 30 over the already impressive current accuracies. This would also help to identify stars that have Earth-like planets, by measuring the wobbles of the stars as their planets orbit them. NASA is considering a program named Kepler that would monitor 100,000 stars to see blips in their brightness caused by the transit of planets in front of them. Also under consideration by NASA is a program named the Terrestrial Planet Finder that will use four satellite-based telescopes to image Earth-like planets. They are looking to detect planetary systems as far away as 50 light years, and to analyze the atmospheric spectra of the planets they find to look for signs of life.

Evidently, targeting remote objects and expecting spacecraft to follow these precise trajectories may be optimistic. The swarm strategy to interstellar clouds reduces this problem by aiming at large numbers of clouds, and by sending large numbers of missions to large targets. Nevertheless, it would be preferable to navigate the vehicles in course. It is also possible that the miniature capsules would be impractical if their biological payloads need substantial shielding during the long interstellar transit. In this case, a larger vehicle can be launched and steered by robots to transport the biological capsules, and to release them when they arrive at desirable targets. It may also be necessary to devise missions where the microbial payload can defrost and multiply/recycle periodically, say every 1×10^5 years, for renewal against radiation-induced genetic degradation. Robotics that could accomplish these tasks is advancing rapidly.

In terms of the payload, biotechnology that can create hardy and versatile microorganisms is advancing rapidly.

On the highest technological level, human interstellar travel can promote life. For example, Oort-belt cometary nuclei can be converted to habitats with resources to sustain up to 10^{13} kg of biomass each, or a human population up to 10^{11} individuals. The orbits of comets can be perturbed to leave the solar system. Human interstellar travel may require centuries of far-reaching developments, including the bioengineering of space-adapted, science-based "homo spasciense". Adaptation to space may also need man/machine composites. This will risk robot takeover but strong safeguards can ensure that control remains in organic brains that wish to perpetuate their, and with that our own, genetic heritage.

We are on our way to becoming a force of life in nature. Our reward will be the legacy of a universe filled with our family of life, which will give human existence a cosmic meaning.

The panspermia papers A3.3, A3.4 and A3.5 in the Appendix give further technical details and references.

4

GALACTIC ECOLOGY

AND

COSMIC ECOLOGY

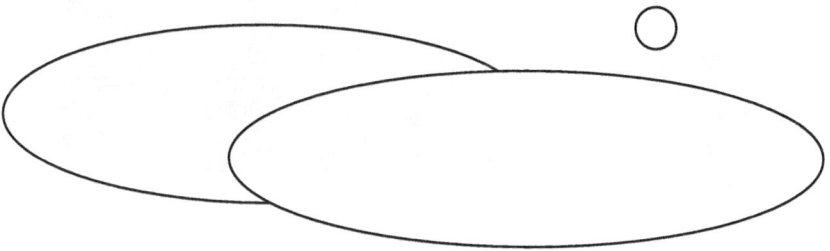

Chapter 4.1 Galactic Ecology and Cosmic Ecology

Our panspermia missions, and human expansion in space, can seed the galaxy with life. What are the prospects for life as it expands into the future?

Life is a self-propagating process of structured matter, and this process always requires a flow of matter and energy. The rules of ecology are universal. In order to predict the future scope of life, we can form some projections of future resources of matter and energy.

We can assume that life will derive energy from the stars. Some of the present stars may last for trillions of years, and more stars will continue to form. We can estimate their energy output through the coming eons. We also know how much energy biomass requires. Combining these data, we can estimate the biomass that can be sustained and how long life can last, about future stars. The numbers are immense.

Future cosmology may be determined by black holes, dark energy and dark matter, whose behavior is not well understood yet. However, our knowledge is growing and we can already apply astroecology considerations to future life and its resources.

We can predict the future scope of life in the galaxy and in the universe using current cosmology. This projected future is immense: up to 10^{34} kg-years of life in the Solar System, 10^{46} kg/years about stars in the galaxy, and up to 10^{59} kg-years in the universe if all matter was turned into biomass. We can express the potential biomass life in numbers, but these amounts, and the richness of life that it can create, are beyond comprehension. The potential scope of future life can strengthen our purpose to expand life in the universe.

Chapter 4.2 Biomass and Populations in the Cosmological Future

Populations Around Red Giant and White Dwarf Suns

In preceding sections we saw predictions about potential life in the future Solar System. Predictions become less certain the more distant is the future. Past observations cover only 14 billion years. The detailed evolution of the Sun, the galaxy and the universe for trillions of years to come can be predicted only by unverified theories. Little research has been done even in theory about the detailed future of the Solar System and its resources, or on the future of other stars and their planets on long time-scales. This section will rely on the cosmological model presented by Adams and Laughlin (1999). Apart from resources, another important factor in the long term is wastage. Recognizing that wastage may severely limit the amount of living matter, future societies with advanced technologies may prevent wastage completely, and this will be assumed in the following discussion.

After the current phase, the Sun will become a red giant and then a white dwarf star. The Earth will be destroyed during the red giant phase but the Solar System itself may remain habitable much longer, even by contemporary humans, during the 10^{20} year white dwarf phase of the Sun (Adams and Laughlin, 1999). The population will only need to move closer or further from the Sun as its luminosity varies. Adjustments to the orbits of space colonies and of free-living humans in space as the Sun's luminosity changes should be simple. The other estimated 10^{12} white dwarfs in the galaxy may similarly sustain life for 10^{20} years.

During the red giant phase the luminosity of the Sun will increase on the order of 2,000 times compared with its current value. This phase will end in an unstable period accompanied by thermal pulses, when the luminosity may reach about 6,000 times its current value (Ribicky and Denis, 2001). Can life find habitable zones in the Solar System during this period?

The habitable zone about a star is defined by the temperatures at which life can survive. For example, it may be defined as the zone where the equilibrium temperature of an object is between 0 and 100 degrees centigrade where liquid water can exist under the pressure of one atmosphere. We may also define a "comfort zone" about the distance from the Sun where the equilibrium temperature is a mild 25 degrees centigrade.

The temperature of an object in orbit about the Sun is determined by equilibrium between the absorption of solar radiation and the emission of thermal radiation. The equilibrium temperature increases with the luminosity of the Sun and it decreases with the heliocentric distance, as shown by equation (A14) in the Appendix 2.3. When the luminosity of the Sun increases to 2,000 times its present value, the habitable zone will be at a distance between 21 and 39 au, and the comfort zone of 25 C will be at about 33 au, about the distance of Neptune. At a later stage when the luminosity of the Sun increases to 6,000 times of its present value, the habitable zone will be between 36 au, (i.e., the distance from the sun to Pluto) and 68 au. The comfort zone will be about 57 au. By this time the Sun will have lost some of its mass, reducing its gravitational pull, and the planets will move out further (Ribicky and Denis, 2001). Neptune, Pluto and the inner Kuiper Belt comets may move to orbits closer to or inside the habitable zones. Space colonies or free-sailing humans can also move to these zones.

The location of the habitable zone will force populations to move to the area of the Kuiper Belt where there are major resources. There are an estimated 35,000 Kuiper Belt objects including Pluto with radii larger than 100 km (Lewis, 1997). The total mass of these comets is at least 10^{24} kg, a hundred times more than the mass of the carbonaceous asteroids. These cometary nuclei are rich in water, organics and inorganic nutrients. The biomass that they can support using all of their elemental contents is 6×10^{22} kg, or in round numbers on the order of 10^{23} kg. A population can live on these resources for a period on the order of a billion years during the red giant phase of the Sun, giving an integrated $BIOTA_{int}$ on the order of 10^{32} kg-years. In comparison, the amount of life that has existed to the present may be estimated as 10^{24} kg-years.

The Kuiper Belt resources will be available only if these cometary nuclei survive the red giant phase of the Sun. Pluto and the other Kuiper Belt objects contain various ices that could evaporate if heated. However, objects that remain below 150 K may not evaporate except for losing some highly volatile substances from their surfaces. This process forms a non-volatile protective crust that prevents further losses, as found in burnt-out comets. Even at the hottest stage of the Sun, objects further than 450 au will remain at these low temperatures. This applies to the outer Kuiper Belt that may extend to 1,000 au. Of course the Oort cloud comets at 40,000 au will be much colder, at 11 K even when the Sun is the most luminous. If most of the Kuiper Belt and the Oort Cloud comets are preserved, the mass will be on the order of 10^{24} - 10^{26} kg including organics and water.

Can these resources be accessed by humans? A velocity of 10^{-4} c can be reached by current solar sails, and this speed is also similar to the orbital speed of the Earth and other objects. Space travel at this speed to the comfort zone at 100 au when the Sun will be most luminous will last only 16 years, and travel to 1000 au to collect resources of the Kuiper Belt will last 160 years. Such travel times are accessible for humans with natural or moderately extended life spans. It is comforting that human populations may survive in the Solar System in this manner through its hottest days, since alternatives such as travel to the colder Oort cloud at 40,000 au would last over 6,000 years and interstellar travel to nearby habitable stars may last up to millions of years, and their feasibility for humans is uncertain (Mauldin, 1992).

Once past the red giant period, biological life, possibly including human life, may continue in this Solar System for up to 10^{20} years using the power output of the white dwarf Sun. It was estimated that at this stage the Sun will be reduced to about the size of the Earth with a surface temperature of 63 K and a luminosity of 10^{15} Watts (compared with the Sun's current luminosity 3.8×10^{26} Watts), that will be powered by the capture and annihilation of dark matter, if this speculative process actually occurs (Adams and Laughlin, 1999). Populations can then move close to the white dwarf Sun and capture its power in a Dyson Sphere (Dyson, 1979b and 1988) as suggested by Adams and Laughlin (1999). It is possible that radiation may be focussed by mirrors or converted to electrical energy for heating to create biologically habitable environments, although at the cost of considerable wastage of energy.

At this stage power, rather than matter, may limit the viable biomass. Assuming a power requirement of 100 Watt/kg biomass, this white dwarf star can support a biomass of 10^{13} kg, possibly in the form of 10^{11} self-sufficient humans. The material for this biomass can be obtained from an asteroid or comet of 10^{15} kg with a radius of about 6 km. This is just one comet of the billions that exist at the present, some of which will survive the red giant phase of the Sun. Although some of the comets will be dispersed by passing stars, (Adams and Laughlin, 1999) a fraction will stay in the Solar System naturally. This may be also secured by technology by moving the comets closer to the Sun or by processing them into a Dyson Sphere (Dyson, 1979b and 1988).

This amount of population may have to be reduced by orders of magnitude if it captures only part of the stellar energy, or if they convert it with a low efficiency, or if there is a need for supporting industry or biomass. Using one percent of the stellar power, a population of one billion could each have access to 10 kW of power at living standards comparable to current industrial societies. The 5×10^{10} kg biomass of this population can be constructed using a small 10^{12} kg asteroid or comet of 620 m radius.

If the low-temperature radiation of the star can be converted to habitable temperatures, lifestyles in this distant future can be Earth-like. From the point of view of the total integrated amount of life, the determining factors are the luminosity of the white dwarf Sun that can support 10^{13} kg biomass, possibly consisting of 10^{11} self-sufficient humans during its lifetime of 10^{20} years. This yields a time-integrated $BIOTA_{int}$ of 10^{33} kg-years possibly consisting of 10^{31} human-years of self-supporting humans or 10^{30} human-years of humans each with a supporting biomass of 1,000 kg. By this scenario humans and a diverse biota can exist in our Solar System for an immensely long hundred million trillion years. The time-integrated biomass is a billion times more than the amount of life that has existed to date, and its time-span is more than ten billion times longer than of life on Earth to date.

These considerations assumed no wastage. However, the biomass constructed using the material resources would usually dissipate a fraction of its mass every year as wastage. On a long time-scale, even a minute rate of wastage can dissipate a very large amount of materials, and the resources must be able to cover this wastage. If the resources are not sufficient, either the rate of wastage per unit biomass, or the amount of biomass, or both, would have to be reduced.

Quantitatively, the relation is given by the equation $M_{biomass}k_{waste}t = M_{resource}c_{x(limiting), resource}/c_{x(limiting),biomass}$. The terms are defined in Appendix 2.1. The concentration of the limiting resource in the cometary materials, $c_{x(limiting), resource}$ is 60 g/kg for nitrogen. As an example, the 10^{26} kg asteroid materials may need to sustain the biota and its wastage about the white dwarf Sun for 10^{20} years. The sustainable rate of wastage $k_{waste} \times M_{biomass}$ is then 60,000 kg/year. If the 10^{15} Watt power output of the white dwarf Sun is all used, it can sustain 10^{13} kilogram of steady-state biomass, consisting of a population of 10^{11} self-sufficient humans. In this case, the rate of waste must be as small as 6×10^{-9} per year, i.e., a fraction wasted per year must be smaller than six billionth of the biomass. On the other hand, if a more realistic rate of waste of say 10^{-4} y^{-1} is assumed, then the steady-state $M_{biomass}$ must be reduced to 6×10^{8} kg, consisting of a human population of at most ten million individuals. Compared with what can be sustained by the biomass and population by the energy of the star, the biomass and population must be reduced by a factor over ten thousand due to even this modest rate of wastage. Supplementary materials from outside the Solar System would be needed to maintain the much larger amount of life that the energy of the star can support. Similar considerations apply to other stars in the galaxy.

Resources and Population in Future Periods of Cosmology

The future evolution of the universe depends on dark matter and dark energy whose natures are largely unknown. Assuming a proton half-life of 10^{37} years (Adams and Laughlin, 1999), the last proton of the 10^{41} kg of baryonic matter in the galaxy will decay after 1.6×10^{39} years and of the 10^{52} kg matter in the universe, after a slightly longer 1.8×10^{39} years. Other loss processes such as the incorporation of matter into black holes, or annihilation by dark matter, may also be possible. In comparison, the observable history since the Big Bang is one part in 10^{29} of these time scales of 10^{39} years. The current age of ordinary matter is approximately the same fraction of its future as the first 4.4 picoseconds (4.4×10^{-12} seconds) in the past history of the universe. Evidently, the future behavior of matter, of the physical constants, and dark matter and dark energy cannot be predicted on the basis of this minute fraction of the future history.

If the universe expands constantly, biology will end but some abstract cognitive "life" may last forever (Dyson, 1979a). If the expansion rate accelerates as it appears to be happening now, even this will not be possible (Dyson, 2001). The present discussions concern the future of biological life only, quantified in the context of the cosmology of Adams and Laughlin (1999).

The expansion of life may depend on human endeavors. If life can survive in our Solar System around the white dwarf Sun for trillions of years, will humans still want to colonize other stars in the galaxy? Societies motivated by panbiotic ethics will strive to do so. In fact, life-centered ethics will help societies to survive, and societies that survive that long would evolve to value life. A purposeful civilization can then colonize the galaxy in a billion years, starting from this Solar System maybe, when impelled by the Sun at its red giant phase.

There will be several types of stars that can support and expand life. Their contributions to the total amount of life in the galaxy $BIOTA_{int, galaxy}$ can be calculated using the equation: $BIOTA_{int, galaxy} = M_{biomass, star} t_{lifetime} n_{stars}$. Here $M_{biomass, star}$ is the steady-state biomass about a given star that can be sustained by the material or energy resources, $t_{lifetime}$ is the lifetime during which this biomass exists and n_{star} is the number of the given type of stars in the galaxy. Note that the product $M_{biomass, star} t_{lifetime}$ is the integrated biomass $BIOTA_{int, star}$ about the given type of star. Of course, these terms may be uncertain by several orders of magnitude. In addition, each type of star will have a wide distribution of masses, luminosities, material resources and lifetimes, and therefore a wide distribution of the biomass and time-integrated biota that they can support. Future calculations must account for these distributions.

The first familiar environment will be red dwarf stars, with luminosities of 0.001 to 0.0001 times that of the Sun, that is, on the order of 10^{23} Watts, which can support 10^{21} kg of biomass. This biomass can be constructed from planets, asteroids or comets if they can be found about red dwarf stars. If these resources are scarce, or if the rate of wastage is significant, then the materials may limit the biomass. Given enough materials, the sustainable integrated $BIOTA_{int}$ with a life-time of 10^{13} years about a red dwarf is then on the order of 10^{34} kg-years, and 10^{12} red dwarfs in the galaxy will allow a $BIOTA_{int}$ of 10^{46} kg-years.

Brown dwarfs may also accommodate life. These small stars are lighter than 0.08 M_{Sun} but somewhat heavier than the gas planets. They radiate heat slowly due to gravitational contraction with a typical luminosity of 10^{20} Watts, which can support 10^{18} kg of biomass for the 10^{10} year lifetime of these stars. This contributes 10^{28} kg-years of potential integrated biomass per star, and the 10^{11} such stars in the galaxy contribute 10^{39} kg-years of integrated biomass.

In the long term, collisions between brown dwarfs will give rise to the last red dwarf stars, possibly with habitable planets (Adams and Laughlin, 1999). The total power output of these last stars in the galaxy will be similar to that of a single star like the Sun, on the order of 10^{26} Watts (Adams and Laughlin, 1999) supporting a total of 10^{24} kg biomass. With lifetimes of 10^{14} years they can contribute 10^{38} kg-years to the integrated biomass in the galaxy. The amount is small compared with other ecosystems, but the significance is that this mode of star formation will produce liveable environments for a long time.

The longest lasting stars in the galaxy will be the white dwarfs. Most of the stars that have ever formed in the galaxy will end up at this stage, yielding on the order of a trillion, 10^{12} such stars (Adams and Laughlin,

1999). As discussed above for our Sun, each can yield a $BIOTA_{int}$ of 10^{33} kg-years giving a $BIOTA_{int}$ of 10^{45} kg-years in the galaxy.

The estimates of biomass for each ecosystem in the galaxy can be extended to the universe by multiplying it by the estimated 10^{11} galaxies. Unless there is local life in these galaxies, they must be reached by colonizing life-forms while they are within the accessible event horizon. If we reach them, biology and human life may exist in the galaxies after they become in effect separated universes. Life may have originated in these galaxies, or life must reach them before they become unreachable because of the expansion of the universe. Our family of life, even branches of humankind, may then exist in billions of universes that will be permanently separated from each other, beyond communication.

These calculations concern upper limits of biomass and populations, i.e., the carrying capacities of the ecosystems as determined by resources of mass or energy. In fact, the populations that will be realized may be limited by the mechanism and rate of expansion of life. Adams and Laughlin (1999) suggested that a natural "random walk" mechanism to populate the galaxy would take trillions of years, while purposeful colonization may succeed in a billion years. Until the difficulties of human interstellar travel are overcome (Mauldin, 1992), life may be spread through directed panspermia, even by current-level technology. The cometary materials in our Solar System are sufficient to seed with microorganisms all the new planetary systems that will form in the galaxy during the next five billion years of the Sun (Mautner and Matloff, 1977 and 1979, Mautner, 1997). The maximum rate of growth of biota in the galaxy, i.e., the biotic potential, is therefore likely to depend on human will and technology rather than on natural limitations.

The above calculations illustrate the immense amounts of potential future life and the considerations that may be applied to quantify these amounts. These estimates are uncertain by orders of magnitude and will be re-evaluated as cosmology advances.

The Ultimate Amounts of Life in the Universe

Finally, we may ask "What is the maximum amount of biological life that the universe could support?" In terms of material resources, the maximum biomass is achieved if all ordinary baryonic matter is converted to elements in their biological proportions, and these elements are then incorporated into biomass. The amount of baryonic matter may be estimated from the mass of the Sun, 10^{30} kg, multiplied by the 10^{11} stars and 10^{11} galaxies, yielding 10^{52} kg. A more sophisticated calculation based on the volume of the universe in an event horizon of 15 billion light-years and the estimated density of baryonic matter of 4.1×10^{-28} kg m^{-3} yielded a similar result of 5.9×10^{51} kg (Wiltshire, 2002).

To maximize Life, all of the 10^{52} kg baryonic matter would be converted to biomass. However, this would not leave any sources of energy, except dark matter, gravitational energy and dark energy and background radiation, which may be impractical to utilize. As a source of energy, a portion of the biomass would have to be converted to energy at a rate that provides the required power for the remaining biomass. The maximum energy may be produced according to $e = mc^2$ by the relativistic conversion of mass. The calculations of the rate of use of matter in this manner are described in Appendix 2.2. To supply a power of 100 Watts per kg biomass, a fraction of 3.5×10^{-8} of the mass must be used per year. The remaining biomass at time t can be calculated from equation (A5) in Appendix 2.2 (substituting k_{waste} by $k_{use} = 3.5 \times 10^{-8}$ year^{-1}). At this rate, all of the 10^{52} kg matter of the universe will be reduced to the 50 kg biomass of the last human after 3.3 billion years, and to the last 10^{-15} kg microorganism after 4.4 billion years. The total integrated BIOTA$_{int}$ will be 3×10^{59} kg-years, possibly in the form of 6×10^{57} human-years. Similarly, if only the 10^{41} kg of baryonic matter in the galaxy is available, it will be reduced to the last human after 2.6 billion years and to the last microorganism after 3.7 billion years. The total integrated BIOTA$_{int}$ will be 3×10^{48} kg-years, possibly in the form of 6×10^{46} human-years.

Converting all the baryonic mass to biomass at the fastest technological rate would achieve the cosmic biotic potential, i.e., accomplish the maximum growth rate of life in he universe. Sustaining the biomass by converting a fraction to energy would in turn achieve the maximum time-integrated biomass in the universe. However, matter for life will have been exhausted and life would become extinct after a few billions of years. This time is much shorter than the time allowed for by proton decay.

As discussed above for the Solar System, the same time-integrated BIOTA$_{int}$ can be achieved with the same resources over a much longer time-span by constructing the biomass at a slower rate and maintaining a smaller steady-state population. For example, assume that we wish to use the matter in the galaxy at a rate that would allow life to exist for the 10^{37} years allowed by proton decay. At the required conversion rate of mass to supply energy, this would allow a steady-state biomass of 3×10^{11} kg, possibly in the form of 6×10^9 humans, yielding over 10^{37} years 3×10^{48} kg-year of time-integrated biomass, possibly as 6×10^{46} human-years, in the galaxy. While life will have existed much longer, the time-integrated biomass remains the same as in the previous scenario of rapid construction and use. Extrapolating to the universe, life could similarly exist for 10^{37} years with a steady-state biomass of 3×10^{22} kg possibly in the form of 6×10^{20} humans. This would yield 3×10^{59} kg-years of time-integrated biomass, possibly as 6×10^{57} human-years, lasting during the 10^{37} years that baryonic matter exists in the universe.

These scenarios describe the maximum amount of biological life that appears possible according to current cosmology. Even if a small fraction of this amount is realized, it will be immensely greater than the amount of life that has existed to the present.

Table 4.1 Estimated resources, biomass and time-integrated biomass (BIOTA$_{int}$) supported by the principal resources of future periods of cosmology.

Location	Materials and mass (kg)	Power (Watts)	Nbr In the Galaxy	Life-time (y)	biomass (kg)[a]	BIOTA$_{int}$ (kg-y)[a]	BIOTA$_{int}$ in galaxy (kg-y)
Solar System	Aster-oids 10^{22}	4×10^{26}	10^{11}	5×10^{9}	5×10^{18} [b] (6×10^{20})[c]	3×10^{28} [b] (3×10^{30})[c]	3×10^{39} [b] (3×10^{41})[c]
Solar System	Comets 10^{26}	4×10^{26}	10^{11}	5×10^{9}	5×10^{22} [b] (6×10^{24})[c]	3×10^{32} [b] (3×10^{34})[c]	3×10^{43} [b] (3×10^{45})[c]
Red Giants	Comets 10^{26}	10^{30}	10^{11}	10^{9}	6×10^{24} [c]	6×10^{33} [c]	10^{44} [c]
White Dwarfs	Comets 10^{26}	10^{15}	10^{12}	10^{20}	10^{13} [d]	10^{33} [d]	10^{45}
Red Dwarfs		10^{23}	10^{12}	10^{13}	10^{21} [d]	10^{34}	10^{46}
Brown Dwarfs		10^{20}	10^{11}	10^{10}	10^{18} [d]	10^{28}	10^{39}
Galaxy	Baryons 10^{41}	mc^2/t			10^{41} [e]		10^{48} [e]
Universe	Baryons 10^{52}	mc^2/t			$<10^{52}$ [f]		10^{59} [f]

Footnotes to Table 4.1

  a. Per solar system.
  b. Biomass obtained using extractable elements in carbonaceous chondrite-like asteroids or comets, based on N and P as limiting nutrients.
  c. Biomass obtained using total elemental contents in carboanceous chondrite-like asteroids or comets, based on N and P as limiting nutrients. It is assumed that by the red giant phase of the Sun the total contents of cometary materials will be available technologically. The numbers are order-of-magnitude estimates, as amounts of resource materials and power are not known accurately.
  d. Biomass based on power supply of 100 Watts/kg as the liming factor.
  e. Per galaxy.
  f. Total in the universe.

Conclusions

In summary, carbonaceous chondrite materials in asteroids and comets are the most likely resources in the Solar System. The limiting nutrients in these resources such as nitrogen or phosphorus may determine the total integrated biomass that can exist during the next five billion year main sequence phase of the Sun. On these time scales wastage is also critical. Minimizing the rate of wastage to 0.01% of the biomass per year would allow a permanent human population of several billions during this period. Further projections are speculative, but biological life may survive the Red Giant phase of the Sun and may continue under the White Dwarf Sun for 10^{20} years. Ultimately, the span of biological life may be limited by the sources of energy and by the proton decay, but these time spans may reach to 10^{37} years.

Current cosmology requires us to re-examine our ethics. In particular, it must be considered that biological life, or even other abstract "life", has finite duration. However, the physics are uncertain, and more solid predictions may require trillions of years of observation. If life is indeed finite, a measure is needed to quantify its amount. Time-integrated biomass was used here. The calculations showed that the potential amounts of life in the cosmological future are immensely greater than life that has existed to date. Our family of organic life can survive in great numbers for immensely long times.

A Life-centered ethics will be necessary to reach that future. With such motivation we can populate space and start new chains of evolution throughout the galaxy. Astroecology and cosmology suggest that these programs can lead to an immense expanse of life that can give human existence cosmic consequences. Our remote descendants may explore if Nature can be transformed in their favor so that Life can exist forever.

5

THE ROAD TO SPACE:

SERVING HUMAN NEEDS

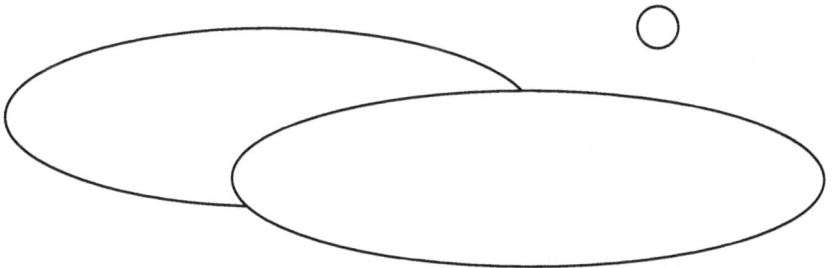

The first steps in space must serve human needs to justify the large investment. Solar energy from satellites, lunar gene-banks, and settlements in space and on planets can serve human needs, house populations up to trillions, help saving the biodiversity on Earth, secure human survival and allow human advancement.

Chapter 5.1 Earth, Space and Human Needs

*H*umans may be altruistic but as all living beings, also necessarily self-centered. The first steps in space will therefore need to serve human needs. The first space applications did in fact provide communications, monitor the environment, and serve political objectives.

Future programs will require substantial human presence in space. After this stage, settlers will increase their populations and expand in space in their own interest.

Future programs in space can provide abundant energy through solar power satellites. If greenhouse warming becomes unavoidable, then space can help to control the climate by solar screens that reduce the energy absorbed from the Sun.

Space can also help to preserve the genetic heritage of endangered species and even of human groups, by establishing permanently frozen genetic depositories in lunar craters.

At the next stage, space colonies, and settlements on the planets, on their moons, and on asteroids, can start to accommodate large populations. This will protect the Earth while allowing large-scale human expansion in affluence, and in the freedom provided by independent civilizations. Dispersed independent habitats will also secure the survival of humankind and of its diverging civilizations.

As space is settled, the merits of spreading life further will become increasingly clear. Settlers adapted to space will be ready to explore interstellar travel for human expansion, and to launch directed panspermia missions aimed at starting life farther in the galaxy.

Chapter 5.2 Satellite Solar Power for Energy and Fresh Water

Demand for power is growing in an increasingly populated and industrialized world. However, fossil fuels will be depleted in a few centuries and they may lead to harmful global warming. Nuclear fission and fusion also have major problems.

A clean space-based alternative was proposed by Peter Glaser in 1968, who suggested collecting solar power in Earth orbit where solar radiation is constant, and to transmit it to Earth by microwaves. Many detailed engineering studies have been done since then and by 2040 Japan intends to build the first Satellite Solar Power Station (SSPS).

Prof. Gerald O'Neill's Princeton group suggested that the structural materials for SSPS stations could be launched economically from the low gravity of the Moon. Human presence on the moon will be required to construct, maintain and operate the SSPS stations and supply the lunar bases. The substantial costs of this program and large profits can be provided by the value of the electric power supplied to Earth. The experience and technology gained in constructing the lunar bases and the confidence gained in living in space can be applied to the cconstruction of larger space colonies.

There is extensive literature on the SSPS program. Here I wish to make one additional suggestion. In the current SSPS designs the energy is beamed to Earth by microwaves and collected by large rectifying antenna stations in deserts or on the ocean. The energy must then be converted to useful forms such as electricity transmitted directly to users, stored as hydrogen for fuel cells or by other means. These steps decrease the efficiency and add significantly to the costs.

There are other applications where these additional costs are minimized. Apart from energy, one of the main problems in the future will be to provide clean water to the increasing world population, especially in the developing countries. Water is already a source of international disputes that may cause wars. Fresh water to alleviate this problem can be produced by desalination but it is expensive.

It may be possible to power desalination plants directly by Solar Power Satellites. The microwave frequency that transmits the energy from the satellites can be chosen to heat and boil or evaporate seawater directly, or the microwaves can be converted to this frequency. The steam can then be used to drive engines to generate electricity or the heat from the condensation vapor can do so. The condensed steam or vapors can be transported as a source of fresh water. The co-generation of energy and water can make the program more economical. A brief technical description and economic analysis at current costs can be made as follows.

The SSPS designs call for stations of 20,000 tons in geosynchronous orbit at an altitude of 36,000 km. The station would provide 5 - 10 GW of power (5 - 10×10^9 Watts), and transmit the power at microwave frequency of 2.45 GHz, similar to the frequency used in microwave ovens. The maximum intensity in the center of the beam would be somewhat higher than the allowed leakage from microwave ovens, but short exposures will not be dangerous to humans or wildlife.

The heat of vaporization of water is 44 kJoules/mol, which translates to 0.68 kW-hours/liter. Therefore a 10 GW station can vaporize 1.5×10^7 liters of water per hour or 1.3×10^{11} liters per year. If a person uses 100 liters of water per day, including agricultural and industrial uses, this station could provide water for 3.6 million people.

The cost of SSPS station was estimated as US$17 billion in year 2000 dollars. Assume that 10% interest/year must be covered at the rate of US$1.7 billion/year. This would yield fresh water at the rate of US$0.013/liter and the cost would amount to US$475 per person per year. Another estimate can be based on the estimated cost of power from SSPS at US$0.20/kW-hour, which would yield a cost of US$0.12/liter, which would imply a high cost of water of US$4,838 per person per year. These estimates do not take into account an efficiency factor and maintenance and transport that increase the costs, but also do not include benefits such as the co-generation of energy that can reduce the costs. Another benefit is of course the supply of clean energy that eliminates the environmental costs of other sources. A more detailed analysis is needed but if the estimates are correct even approximately, the program seems to be economical. It will become even more economical when the costs of launching payloads to space are reduced, and when other sources of energy and water decrease in coming years.

Chapter 5.3 Mining the Asteroids

Beyond energy, other resources that may run short on Earth are materials such as metals. Some types of asteroids contain almost pure nickel and iron that have significant commercial values. It may be economical to tow these asteroids into Earth orbit and then transport them in pieces gradually on Earth. They may be also used for manufacturing while still in orbit. The potential mining of asteroids was studied in conjunction with the space colonization programs summarized below, and was reviewed recently in a book by John Lewis.

The book discusses in particular the metal resources in the asteroids. It estimates that the value of iron in the asteroids is US$3.5x10^{19}, which would yield 5 billion dollars to each person on Earth (of course, the value per ton would decrease from $50 to a few cents with such a supply). More realistically, a 2 km radius iron-nickel asteroid would weigh about 2x10^{14} kg and would be worth on the order of $10^{13}, which is comparable to the total annual US gross national product. It could provide the world with enough steel for one thousand years at the current rate, or to put it another way, a small asteroid with a 340 m radius weighing 10^{12} kg could provide all the steel and many other metals for the world for a year. The other metals in the asteroid may include about 10% nickel, and enough platinum, gold, silver, copper and other metals which would increase the value of the asteroid about ten-fold. It would be economical to retrieve such an asteroid even at the cost of $1,000 billion, much below the likely cost of a space program of this magnitude. Such an operation would require a large crew in space. Assuming the current launch costs of $10,000/kg, the metals from this small asteroid would pay for launching 10^{8} kg of humans and supplies. If each astronaut requires 10,000 kg of supplies and support materials, this would still allow launching a colony with 10,000 residents. Therefore, the retrieval of one small asteroid could pay for the construction of the first space colony, which can bootstrap exponentially to the large populations discussed in other sections. Of course, a much smaller crew would suffice for the retrieval program, which could therefore engender a large profit. While such a project is unlikely in the near future, there may well be a greater motivation when the resources on Earth become fewer.

Carbonaceous Chondrite Asteroids: Resources of Organics, Nutrients and Water For Past And Future Life. On June 27, 1997 the

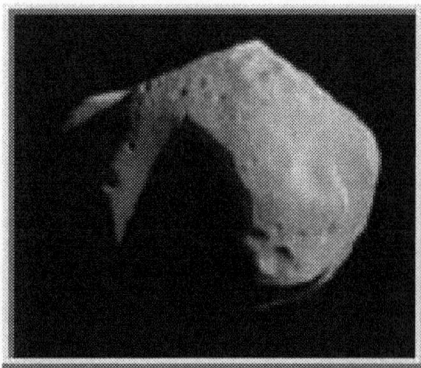

Near Earth Asteroid Rendezvous spacecraft flew past asteroid 253 Mathilde, one of the blackest objects in the solar system. Mathilde is 52 kilometers in diameter. On June 27, 1997 NEAR made a spectacular 25 minute flyby of Mathilde which resulted in more than 500 images of a dark, crater-battered little world that dates to the beginning of the solar system. Mathilde is classified as a carbon rich C-type asteroid which could supply organics and soil nutrients to space colonies.

Chapter 5.4 Space-Based Gene Banks

Population growth and the destruction of natural habitats are leading to the rapid extinction of many species. Genetically distinct human groups are also disappearing before their genetic heritage is explored.

Moving populations to space can solve these problems, but it is unlikely to have a major impact for centuries. After that, the Earth can be restored to a low-population reserve with a large biodiversity. The species that were lost in the interim can then be restored by cloning, a technology that is maturing rapidly. To re-clone endangered species, their genetic material must be preserved. Space can contribute by providing secure storage in deep cold.

For this purpose, the genetic materials of endangered species must be maintained permanently near 77 degrees Kelvin, near liquid nitrogen temperatures, or lower. Human institutions would have to maintain these storage facilities for centuries or millenniums. However, institutions are unstable because of social, political and economic changes, including wars and terrorism. Permanent safety is best assured at storage sites that maintain the needed low temperatures naturally and that provide barriers to access. The required conditions are available in permanently shaded polar lunar craters with equilibrium temperatures of 40 - 80 K, on the moons of Saturn, and in storage satellites. A genetic depository can be incorporated readily into planned lunar programs.

The Current State of Extinctions and Cryoconservation

Over the next 50 years, 15-30% of the estimated 5-10 million species may disappear due to human pressures. The losses include hundreds of vertebrate, hundreds of thousands of plants and over a million insect species. The gene pools of many human ethnic groups are also threatened. For many animals, adequate conservation of habitats is unfeasible and active breeding programs cover only 175 of the many thousands of species threatened. The genetic heritage of the living world, accumulated during aeons of evolution, is being wasted in a short period.

Against such losses, scientists are starting cryopreservation programs of genetic material in tissue samples, semen and embryos. However, funding is already tenuous. Over the centuries, war, sabotage, disasters, economic depression or just a loss of interest are bound to happen. Even a brief disruption of the permanent refrigeration can destroy the samples. Secure preservation requires remote sites immune to such disruptions, at locations where the samples can survive periodic abandonment and remain stable indefinitely at natural temperatures around 80 K.

Finding Suitable Sites

No terrestrial sites satisfy these requirements. However, the required equilibrium temperatures exist at lunar polar craters, or at more remote planetary sites that can be used for long-term backup storage. About 0.5% of the total lunar surface area is estimated to be permanently shadowed and remain below 100 K. Thermal balance calculations suggest steady-state temperatures of 40 - 60 K at the shadowed centers of polar craters. The lunar sites are safe from earthquakes due to the low lunar seismic activity, with an average of five events per year with Richter magnitudes of only $2.2 < m < 4.8$. At these sites, burial in one meter of regolith soil will protect the genetic material against the solar wind, solar and galactic cosmic rays and micrometeorite impact.

Other planetary locations with equilibrium temperatures of about 70-90 K exist on the moons of Saturn. The atmosphere of Titan with a surface pressure of 1.5 bars can protect the genetic material from the vacuum of space, but prevents easy access and retrieval. Ease of retrieval favors small, low-gravity moons such as the high-albedo Enceladus with an average temperature of 70 K, or Phoebe, whose large distance from Saturn (1.3×10^5 km) and small size (220 km diameter) allows a small escape velocity.

Feasible deposits may also be established on storage satellites in Earth orbit. Shields for thermal insulation and from ionizing radiation can be provided. Simpler and more secure, but harder to retrieve, are storage satellites that may be deposited in solar orbit at about 10 au where equilibrium temperatures are around 80 K.

On lunar or planetary sites, samples will be subject to space vacuum with at most 2×10^5 particles/cm^3 in the lunar environment. Although some microbial materials can survive space vacuum and low temperatures for several months, tissues from higher plants and animals should be encapsulated against dehydration that can cause biological changes. Cryopreservation can be used without extracting DNA by preserving tissue samples. Whole cells or small tissue samples will be easier to prepare for conservation, and to clone back later. To preserve microbial fauna, it is not necessary to identify each species. Microbial

110

filtrates from various habitats such as soils, lakes, oceans and those living on other organisms, can be preserved.

Requirements for Restoring Endangered Species

To restore an endangered species with viable genetic diversity, material from 20-30 unrelated individuals is regarded as the minimum, and 0.1 g material from each founder individual can be sufficient to clone an individual using advanced genetic techniques.

The germplasm or tissue samples of one million endangered species can be accommodated in a payload of 2,000 kg. At expected future launch costs of $1,000 - $10,000/kg, the genetic heritage of one million endangered species can be stored in permanent safety at the cost of $2 to $20 million, or only a few dollars per species. Secure permanent storage for centuries or millennia can be achieved at smaller overall costs than storing samples in vulnerable terrestrial storage for a few decades.

Lunar sites appear to have the best combination of required physical conditions, feasibility and security features. The required transport of 2,000 kg is comparable to the proposed Lunar Cluster Telescope and less than the Large Lunar Telescope or the Lunar Synthesis Array. Constructing an underground vault of 10 m^3 is much less demanding than constructing lunar observatories or permanent lunar bases. Therefore, the genetic storage program can be incorporated into other lunar programs at little extra cost. Such missions are likely within 20 - 40 years, which allows time to collect the genetic material.

Even before the permanent storage sites are constructed, payloads of the genetic materials can be soft-landed into the cold lunar craters. They would be stable there, or may be deposited into shallow burrows by robots. When humans return to the Moon, these collections can be transferred to permanent depositories.

In addition to plant and animal species, the genetic material of human groups is also being lost as populations disperse and intermix. This material is important for the understanding of human evolution, history and biology. Genetic materials of vanishing human groups can be also preserved in space-based depositories.

The Philosophy of Space-Based Cryoconservation of Endangered Species

Conservation of endangered species is a popular program. Cryoconservation can therefore add an important ethical component and help raise public support for the space program. Conversely, the permanent safety of the genetic material can also make cryoconservation more attractive and fundable. Many nations may wish to participate to secure the genetic heritage of their unique biota and of their ethnic groups.

The genetic heritage unites and enriches the living world, and preserving it is a top ethical priority. Until habitat losses are controlled, cryoconservation may provide the best change to secure and eventually revive many endangered species. For this purpose, space-based depositories can provide the most cost-effective and secure means for permanent storage of irreplaceable genetic materials.

A further benefit is the information that we will gain about adapting life to extreme environments. Many present species have developed adaptations to extreme temperatures, to dry or wet conditions, to UV and other types of radiation, to high pressures under the ocean, to extreme salinity and pH conditions. All of these conditions are likely to be encountered in diverse new planetary habitats or in free living in space. The biological information in these species can be used in adapting life to space, and is therefore important to preserve.

The References can be found in the paper in Appendix A3.9.

Chapter 5.5 Controlling the Climate: Solar Screen Against Greenhouse Warming

Intercepting about 3% of the Earth-bound solar radiation in space can prevent the expected greenhouse warming and surface UV incidence. Solar screen units based on thin-film material can be deployed in levitated solar orbit or about the Sun-Earth L1 libration point. The mass requirements are 10^9 - 10^{11} kg which can be obtained from lunar or asteroid sources and processed using vacuum vapor deposition. The screen could prevent the projected $10 trillion economic damage of climatic warming by 1% of the cost. Using screens of dust grains could further reduce the deployment and operation costs by an order of magnitude. Cost/benefit analysis justifies a $4B research program at present and a $100B program for deployment 40 years prior to the onset of serious climatic warming. Most importantly, such screens should be studied as a measure of last resort if the Earth becomes threatened by runaway greenhouse warming due to positive feedback effects.

It may become necessary to control the climate of the Earth or the planets in order to keep, or render, the environment habitable. For example, the climate of the Earth is expected to change because of increasing levels of CO_2 and other greenhouse gases in the atmosphere. The resulting greenhouse effect is likely to lead to climatic warming by 1 - 5 C, with flooding, droughts, wildlife extinction, and economic damage on a global scale [1 - 3]. These problems are compounded by increased UV radiation on the Earth's surface because of the depletion of the ozone layer. The Antarctic ozone hole is already increasing UV exposures in the South Pacific regions of Australia and New Zealand.

Environmental solutions to these problems are desirable but costly. For example, developing renewable sources for the world's energy needs may cost trillions of dollars [4-6]. If environmental solutions fail, the damage by the greenhouse effect would also amount to trillions of dollars [5].

We recently pointed out that a climatic crisis may be averted by decreasing the solar radiation incident on the Earth in the first place [7]. This may be achieved using existing or near-future space technology, with expenses much less than the cost of greenhouse damage. However, such intervention should be done cautiously since the ultimate effects of climatic intervention may be hard to predict [5].

As intervention with the climate is becoming feasible, it should be seriously studied and debated. Designing the space-based solar screen is an interdisciplinary project involving astronomy, astronautics, optics, material science, geophysics, atmospheric science, climatic modeling, ecology, economy and international law. This paper will present some general considerations in several of these areas, and survey specific research needs.

Reprtints of papers on this subject are presented in Appendix A3.3, A3.4 and A3.5

Chapter 5.6 Space Colonies and Planetary Terraforming

The increasing world population is already approaching the carrying capacity of the Earth. The scarcity of land and resources aggravates territoriality and tribalism, which, combined with mass weapons, can threaten human survival. Furthermore, populations will skyrocket further as medicine progresses, and especially if aging is stopped. Populations may then be limited by forced birth control under tyranny, or nature may intervene with diseases, wars and famine.

But human growth need not be limited if we can access more resources and create more living space. This can be accomplished by the construction of space colonies, each housing millions in comfort, similar to large cities or small countries. The program was suggested and explored in engineering detail by Prof. O'Neill in the 1970's. The main point of space colonization is that the growth of colonies is exponential. The first colony can build several others, each of which can then build yet several more. The vast human populations that were described in the Astroecology sections of this book can be accommodated in space colonies within a few centuries.

The first colonies will be built in the vicinity of Earth with structural materials launched easily from the low gravity of the Moon. As noted in the Astroecology sections, the colonies can obtain organic soils and water from the carbonaceous asteroids located mostly between Mars and Jupiter. It may be advantageous to build future colonies in this zone to take advantage of the nearby resources. Mars can then also serve as a major local planetary center. By the time that space colonies are constructed in the asteroid belt, Mars may be ready to fulfil this role. There are many proposals to terraform Mars, which could be accomplished in a few centuries. Settling Mars and the asteroid belt is also desirable in the long term since they will be the first refuges when the Sun starts to expand.

Space cities have a many good features. First, the colonies can use the abundant and permanent solar power available to them. Agriculture can use individual capsules with conditions optimized for each crop, and unlimited growing seasons. Industrial areas can be located in separate capsules, allowing pure environments in the colonies. Industry can make use of conventional gravity environments created by rotation, or reduced gravity to create special crystals, pharmaceuticals and alloys. The climate and the lengths of day and night can be controlled by the residents. Disabled people can live in reduced-gravity zones where it is easier to move about.

Each colony will be owned by its residents and there will be no "ancestral homelands" to fight over or neighbors to clash with, and neighboring dictators who may try to take over. For any society that wants more territory, it will be far easier to construct new colonies than to take over existing ones. Space colonization will promote peace.

An important advantage is that large numbers of independent colonies will allow a wide range of social diversity without friction amongst different civilizations.

In particular, the size of the colonies will allow for self-governance by pure direct democracy. Each colony can choose its own laws and culture. Indeed, it can be argued that direct democracy distils out from the diverging will of individuals the desire common to all, which is respect for life and human survival.

Indeed, it can be shown by mathematical argument that the communal will is wiser than individual leaders. For example, assume that decision "a" is better than decision "b". Let us assume that a wise government is likely to chose "a" with a 99% chance, while an average, barely intelligent voter chooses "a" only with a chance of 51%, barely better than randomly. Even so, statistics shows that the probability of the majority vote choosing "a" increases and approaches 100% as the number of votes increases. In other words, at the end, the communal will is always wiser than any individual ruler or government.

A program of space colonization can satisfy the most important human needs. Once based in space, humans will adjust biologically through designed evolution.They will live freely in space, using asteroids, comets and space habitats only for refueling. The results of the astroecology experiments described above show that asteroid resources can sustain populations of trillions. Expanding life in space will be self-evident for these space-born populations. Both culturally and by easy access to space, they will be well placed to start panspermia programs to seed the galaxy, to move on to new solar systems and to carry on the seeds of Life within them.

Chapter 5.7 Life in Space to Serve Human Needs

Human society, even human biology, populations and ecology, are entering a new era that is controlled by human actions. These developments promise the ultimate prize of expanding life and human existence and securing them for the habitable lifetime of the universe. They also incur the ultimate danger of destroying our species and with it the future of all life.

It is therefore vital that we proceed to expand and secure life in space as fast as possible. However human nature is self-centered, and space is far from the personal concerns of most people. The move to space requires concerted efforts of large organizations and many people, and the investing of a significant fraction of the wealth of nations. This is feasible politically only if space offers benefits that people could feel on a personal level.

Such benefits are possible, and will be needed soon as the population of the Earth increases and its resources are depleted. Space can fulfil two of the greatest needs, energy and living space. Space can allow us to control the climate of the Earth, and offer safe storage for fragile genetic materials.

This chapter discussed only a few examples of the benefits that may make the first steps into space profitable economically and feasible politically. Once an infrastructure is established, it will be easy both technologically and psychologically for humans born in space to expand their habitats and population.

Space can satisfy our deep territorial desires. Space can allow us to multiply our seed and to safeguard its future as long as the universe allows life. We may hope that they will also appreciate life-centered panbiotic principles and proceed to seed the universe.

Space colonies such as these "O'Neill cylinders" can each house millions. Colonies using materials from the moon and asteroids, and clean, permanent solar energy can eventually accommodate trillions of people. From their space-based position they will construct satellite solar power stations to supply energy to Earth, secure human survival, and accommodate diverse cultures. They can easily launch panspermia missions, and some colonies may embark for other solar systems in the galaxy. Art courtesy of Space Studies Institute.

116

6

THE ULTIMATE FUTURE

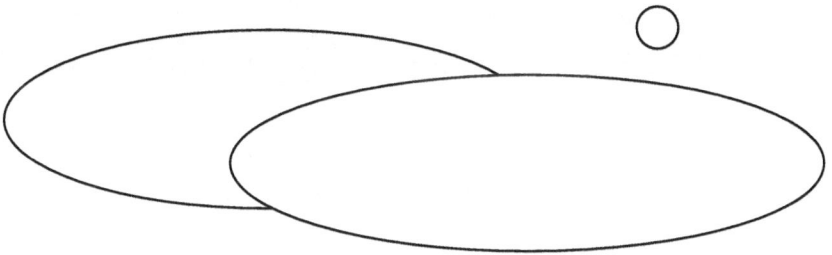

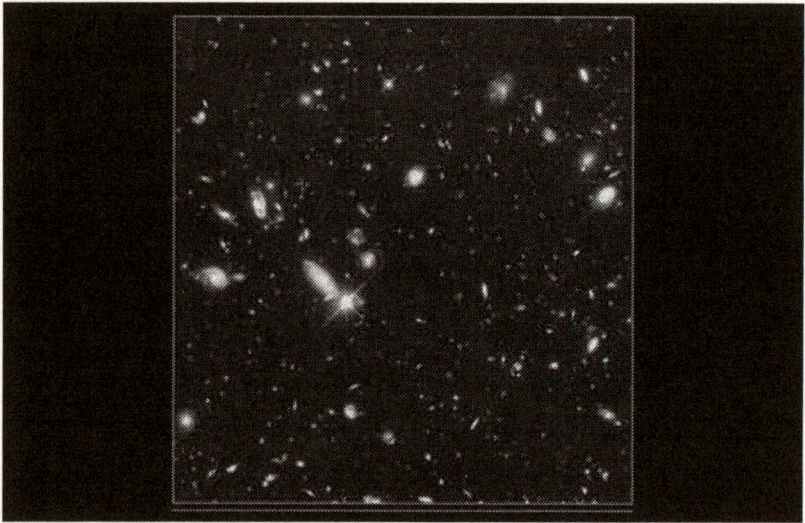

The Hubble Space Telescope Deep Field View shows 1,500 galaxies at various stages of evolution. The 15 billion years of their history is only a minute fraction of the future in which Life can populate the universe.

Chapter 6.1 Our Cosmological Future

We are charged with the future of humankind, and of life. We shall need to adapt to space to survive there. Human brains may grow ever more complex and interface with computers; vision may extend into the ultraviolet and infrared; communication may be achieved by radio or laser organs; sexual reproduction may be replaced by embrio cultures; and aging and death may be replaced by immortality. Gene/protein life may itself be altered by extended DNA bases and synthetic proteins.

Will these developments serve life? This can be secured if control remains with gene/protein brains with a vested interest to continue our gene/protein life-form. To secure suvival, this controlling intelligence must always be guided by a panbiotic ethics that aims to propagate life.

Can we predict and control the future in order to secure life? Human knowledge is constrained by our limited capacity vs. the complexity of nature; by our finite life-times; and by the limits of small and large scales that we can experience. We cannot know if nature is deterministic or if it may follow many paths, since we cannot replay its history. We cannot know what there is, or may be, beyond the limits of our knowledge.

We do know however, that life is unique, that we are part of life and that we are charged to propagate life. We must act on this knowledge, believing that we can shape the future. We must do so since our expectations can be self-fulfilling.

Will life end in the universe? If it will, is there a purpose? If the future is infinite, we cannot observe it; and if it will end, there will be no one to observe it. An open-ended future allows hope and calls for action.

There may be infinity in a succession of expanding universes. Our universe expanded from a volume smaller than an atom that may have been a particle in a previous universe. This preceding particle/universe may have expanded from another yet more compressed epoch and so on back to infinity. Our universe will expand and cool, and each particle may become a separate universe that will cool and expand again, dividing into a new generation of expanding particle/universes, and so on to infinity. This model is beyond experiment, but it has a universal symmetry. We see ourselves in the center between a compressed and dilute universe, and so do the inhabitants of each epoch. Maybe, our descendants can then configure the next epoch so that it can also harbor complex life.

An unknowable future allows hope. Indeed, we must act with

hope, since our projections of the future can be self-fulfilling. It is for us to secure that our descendants will exist. We must institute in them an ethics of life, so that each generation will secure the next. Generation by generation, our family of life will then expand into an immense future. It will be for our distant descendants to seek infinity.

Our cosmological future will be secure if each generation seeks to safeguard and propagate life. Fulfilling the inherent purpose of life, our existence itself will find a cosmic purpose.

Chapter 6.2 Designing Future Humans

Objectives and Principles of Designed Human Evolution

The ultimate human purpose is to safeguard and propagate Life. The objective of designed evolution is to promote this purpose: we wish to design future humans who will be capable and committed to propagate Life.

Such advanced humans must possess:
- Intelligence that is capable to design a future;
- An innate drive for self-propagation;
- A vested interest and lasting will to propagate organic gene/protein Life, and
- The physical ability to pursue these actions.

There are virtually infinite biological forms that could all propagate Life. Some may develop naturally, and some by human design. Both will follow unpredictable paths, and a great divergence of life forms will emerge.

As humans, we don't only wish to continue Life, but also to continue our own genetic heritage, at least in some of the future branches of Life. What needs to be preserved in these species so that they will remain human?

On the most basic level, we are organic cellular beings. Our genomes preserve the shared biological heritage back to the first primeval cells. We share the basic biochemical cycles, the DNA code and the mechanisms of gene translation, the use of ATP for energy, and the mechanisms of membrane transport, with all other living species. To remain part of Life, this beautiful complex creation of nature must be preserved in our descendants.

Although new genes are added in natural evolution, the development of each individual recapitulates the evolution of the species. As the embryo evolves, it re-enacts the development of life from unicellular organisms, and each stage builds on the preceding stage as evolution is retraced. In the future, the human stage will be just another past stage through which our descendants must pass to their stage of evolution. In this manner, our genetic heritage will be preserved in each cell of our descendants, just as we are preserving in our development the genetic heritage of our ancestors.

While the cells retain their gene/protein heritage, our existing organs can be re-designed or new organs can be created to suit new environments. The present human body was produced by adaptation to accidental circumstances such as Earth-normal gravity, a diurnal cycle, and food sources. Not all the features that accommodate these arbitrary conditions are essential for humanness.

However, there are features that when combined make us uniquely human: conscious self-awareness, intelligence, emotions, instincts of survival, the desire for procreation, the love of young life, the appreciation of life, an understanding of the universe, an awe of the grandeur and mystery of Creation, the conscious joy of living, the appreciation of beauty, and the recognition of good and evil. Our successors must retain these attributes to remain human.

These features must be retained even as physiology changes for the ultimate adaptation, to live in free space itself, although probably still about stars and planets as sources of matter and energy. Adapting to free space will affect all the levels of the biological hierarchy, from new biomolecules to cells and organs, organisms, societies, and an entire new galactic ecosystem.

We may be able to affect the future of this evolution, but Life will in any event find ways to explore all possible paths, subject to the tests of survival. As we cannot prove whether or not we can affect the future, we must assume that we can, and act accordingly. We must choose the guiding principles carefully, as our visions of future Life may be self-fulfilling.

The Principal Requirement: Living Brains

Free living in space will require many adaptations: new durable organs, impermeable skins, solar sail wings, organs for radio communication, radiation resistance, and possibly developing a slow-paced but long life span. These adaptations may affect even the basic biochemistry of the cell. Some human cells may need to merge animal and plant traits to perform photosynthesis for self-contained free life in space. New features may require new proteins, possibly including new types of amino acids. The corresponding genetic code may include new nucleic bases or other types of information-containing molecules.

Moreover, organic and inorganic machine organs may be combined. At the extreme, biology can be completely displaced by robots. However, this would defeat the essential purpose of Life, self-propagation. Extinguishing Life in this manner would betray our innate purpose as living beings.

How can we assure that our successors will propagate cellular organic Life? Nature answered this question by giving us the drive for self-propagation. The result of having this drive is that Life itself is propagated with each birth. This drive for self-propagation must itself be propagated to assure that Life will survive. Future beings must always continue to have a vested self-interest to propagate Life.

The future will be controlled by intelligent beings. They will have a vested interest to propagate organic Life only if they are themselves organic gene/protein cellular beings. For this reason, at least the controlling organs of these future beings must remain members of our family of cellular gene/protein life.

Supplemented Biological Brains

The intelligence of organic brains is limited but this is acceptable as long as they propagate Life. After all, Life advanced for four billion years without higher intelligence. Nevertheless, the drive for progress will seek to create superior intelligences.

Should these future brains be interfaced with electronic computers? Doing so can vastly increase brain capacity, but it would help the takeover by inorganic machine computers. Better, advanced brains can be constituted of new designs of biomolecules. For example, protein-based optoelectronic digital computer memories can use bacteriorhodopsin units with alternate switchable states. Computing based on DNA is also being discussed. Such new brains are not based on neurons but can still reproduce through gene/protein mechanisms and would therefore remain legitimate living organs.

For more capacity, individual brains can be interlinked. Examples of dispersed, collective organisms already exist in bacteria such as *Mixococcus* or *Bacillus Subtili*[7] that can exist both as individual cells or function as differentiated multicellular organisms in colonies. On a human level, science already consists of individual minds interlinked through communication. This interlinked body of science is a superior intelligence with a vast knowledge base, and it achieves greater and faster progress than any individual could. Similarly, a vast population of trillions free-living beings in space can interlink to form a super-brain that permeates the Solar System, even the galaxy.

The essence of future humans will remain cellular gene/protein brains with a vested interest to continue the family of organic Life. These brains can possess consciousness, intelligence, emotion and purpose, which will be in fact needed for survival. With this combination of properties, they will retain the essence of being human.

Body Size and Closed-Cycle Metabolism

The minimal requirement to retain organic gene/protein brains allows much freedom in future evolution. Nevertheless, the biological needs of these brains will define other aspects of human physiology and even of astroecology.

The properties of organic brains will evolve, but for now we will assume that a biological brain must retain at least its present size and energy requirements. The brain uses a power supply of about 800 kJoules/day, or about 9 Watts by converting chemical energy stored in glucose and oxygen. In free space, energy will be derived from solar radiation. Given the solar constant of 1,353 W/m^2 at 1 au, and a conversion efficiency of 10%, this will require solar energy incident on about 670 cm^2, i.e., a square collector of 26x26 cm.

Brain cells contain the basic biochemistry of all cells, plus specific features such as neurotransmitters and receptors. Sustaining this chemistry may require a liver-sized chemical organ, blood to transport the chemicals, a heart for circulation, and supporting tissue. Altogether, we may assume a minimum organic body mass of 20 kg and energy demand of 20 W, which can be powered by a solar collector organ of 38x38 cm.

To maximize the population in space, it will require using and wasting minimum possible materials, which is helped by recycling. In particular, the ecological cycle of oxygen and carbon dioxide can be combined in self-sufficient organisms. For example, algae can be embedded in wings that also serve as solar sails, taking up CO_2 and releasing O_2 into the bloodstream. Alternatively, chloroplasts can be incorporated directly into human cells, as happened in evolution when eukaryotes incorporated cyanobacteria and became plant cells. A similar process can incorporate chloroplasts from plants into human cells. When exposed to light as they circulate in he blood through solar sail wings, these cells can manufacture sugars from carbon dioxide that is recycled after respiration. The products can then be delivered to the other organs as a source of materials and energy.

These are only some examples of possible future physiology. The main point is that the needs of organic gene/protein brains will also require supporting gene/protein organs. This further expands the amount future living matter and poses a further barrier to machine takeover.

Trans-Natural Biology: Extended DNA and Proteins

Artificial evolution can mutate DNA and enzymes to produce new materials for space adaptation. These organisms will still belong to Life if their DNA and enzymes are made of the natural constituents.

However, the properties of proteins can be greatly extended by including new synthetic amino acids. For example, phenylalanine was replaced by a fluorinated analogue to produce a fluorinated protein with Teflon-like chemical resistance that could be useful as post-human skin in corrosive environments. Another engineered protein, Pronectin F, which is produced by combining an artificial silk gene with a human gene for fibronectin, could be used to grow thin "wings" for solar sailing. These sail organs can be manoeuvred by synthetic contractile elastin-like poly(VPGVG) protein fibers that are actuated by temperature changes.

Many other modified proteins can be produced by transgenic organisms to extend the survivable range of pH, ionic strength, water activity, temperature and radiation. Nature already provided such capabilities in extremophile microorganisms.

To encode new amino acids, the DNA code will need to be extended. Natural DNA encodes each amino acid using a three-letter codon, each letter chosen from the four natural nucleic bases, allowing different codes for 64 amino acids. New DNA that uses an m-letter code chosen of n bases can code for n^m different amino acids. Changing to a six-letter codon of the four natural bases could increase the choice to 4096 different amino acids. Alternatively, a three-letter codon may be retained, but use additional base pairs. A three-letter codon of eight bases can code for 512 amino acids, and a six-letter codon of eight bases can code for 262,144 different amino acids.

The first approach, using longer codons of the natural nucleic bases seems more efficient because of the exponential relationship. Information bearing polymers may be also constituted of different types of molecules other than DNA.

Does an organism, constituted of artificial proteins and coded by artificial nucleic bases or other artificial biopolymers, still belong to the family of Life? They can be accepted as fellow life as long as they share the principle mechanisms of gene/protein cycles. However, such organisms should not displace post-humans who possess our genomes, if our genetic heritage is to be continued.

Society and Locomotion

Social interactions require communications and transport. Communications within the habitable zone about the Sun will be easy if electromagnetic transmitters and receivers replace the natural organs of speech and hearing.

To estimate the locomotion needs, we assume a population of 10^{16} that inhabits a habitable zone with equilibrium temperatures from $278°$ to $305°$ K that are located between 1.0023 to 1.0500 au, with a volume of 2.1×10^{33} m^3, giving a living volume of 2.1×10^{17} m^3 per individual, or an average distance of 3.7×10^5 m between individuals. By analogy with current city-size communities, an individual may wish to be within easy reach of one million other individuals, occupying 2.1×10^{23} m^3 of nearby space with a radius of 3.7×10^7 m. In present social interactions meetings are usually arranged with a week's lead time that is 1/5,000 of the lifetime of an individual. Assuming life-spans of 10,000 years, social meetings would be scheduled with lead-times of two years, during which each individual would need to travel about 10^7 m to the meeting location, at the modest speed of 0.3 m/sec, or 1.1 km/hour.

For locomotion, solar sailing does not require the expenditure of counter-masses and energy. An organism with a body mass of 20 kg using a 90% reflective sail of 10^{-3} kg m^{-2} of 126 m radius, giving a total effective areal density (total mass/area) of 0.0014 kg m^{-2}, will achieve at 1 au an acceleration of 2.3×10^{-4} ms^{-2}, and will therefore acquire the necessary speed in 22 minutes. Alternatively, accelerating for 36 hours can decrease the travel time to 7 days. For stopping, two travellers from opposite directions can simply collide with each other.

These examples illustrate that solar sailing can provide adequate locomotion for social life. Another possibility is "hopping" between asteroid-based communities allowed by the low escape velocities of asteroids.

Psychology and Reality

A great danger is that future minds may lose touch with reality. Today, this happens when drug addicts seek happiness in hallucinations which are easier to achieve than real fulfilment. Addicts to the fantasy world of television, virtual reality and the internet are other signs of such dangers.

Even intellectual pleasures can be dangerous. There will always remain challenges in the infinite combinations of molecular and mathematical constructs, which can provide endless creativity. Future neuro-psychological engineering should produce brains capable of lasting happiness when searching for understanding and beauty in these abstract domains. This can in fact serve as a motivation to continue life. Nevertheless, this too can be dangerous if intellectual pleasures cause individuals to ignore the real needs of survival.

It will be easy for future beings to design themselves to live in permanent internal happiness. However, life will always need to confront reality, where entropy tends to destroy biological structure. Life fights these disruptive forces by continuing efforts for survival and reproduction, driven by pleasures that that these pursuits cause. In fact, this is why we evolved to find pleasure in food, warmth and sex that serve survival and propagation.

Similarly, in the future as well, pleasure must remain linked to physical survival and propagation rather than purely internal events. Future beings will then find too the ultimate happiness in promoting their own genetic heritage, and with it, the patterns of organic life.

Reproduction

Self-engineered beings may not wish to preserve the awkward and painful ways of natural reproduction. Future humans are likely to be grown in tissue cultures or even to be assembled directly as adults from cultivated organs. These methods will also allow a better control of the products.

The logic of evolution linked reproduction to the most intense physical and emotional pleasures. Will Life survive without the desire of the individual for sex and heirs?

Without these desires, continuing Life will depend entirely on conscious logic. This can be dangerous. Yet even now, raising and educating the young are pursued also for emotional and moral reasons. Humans also cultivate gardens, raise pet animals and protect wildlife for the same reasons. Future survival will depend on preserving the empathy to living beings.

The motivation for advancing Life will lead to self-directed evolution. Every generation will then design successors with features that they perceive as improved. However, the effects of these changes cannot be fully predicted because of the complexities of biology and society. How will successors with unpredictable features in turn design their successors? Their motivations will develop unpredictably. The uncertainty will multiply with succeeding generations, in a process of non-linear evolution.

Designed evolution will be much faster than natural evolution and it will take diverging paths. No finite intelligence can foresee the possible outcomes. On this background, designed evolution is as random as natural mutations. Both lead to changes that are tested by survival. At the end, the Logic of Life, not humans with limited intelligence, will define the course of evolution. The mechanisms of Life may change, but the Logic of Life is immutable and permanent.

The immutable logic of Life will dictate that only species that are dedicated to propagation will endure. Therefore Life-centred panbiotic ethics demands that the will to continue Life will itself, always be preserved. Whatever means future self-designing species will use to reproduce, a panbiotic ethics will be essential for securing the future.

Can Life Survive Immortality?

What will future minds contemplate, living in the expanses of space for eons? First, against deadly tedium, the rate of internal thoughts should be made comparable to the rate of external inputs. A fast thinking brain detached from external input would be like prisoners in isolated dungeons, where the mind turns on itself and ends in madness. One form of this madness may be absorption into fantasy and the loss of reality.

An external danger of immortality is that once all the resources are taken up by immortal individuals, there is no room for progress. Without progress, there is no room for further growth or adaptation to new challenges. This is why Nature denied immortality to any living species.

Of course, the immortality of individuals will block progress only when all the vast resources of space are bound up. This however, can happen quickly as Life can colonize the galaxy in a few million years, a small fraction of at least trillions of habitable years in the universe. If individuals insist on immortality, then the progress of Life to ever higher forms will have to cease at some point.

However, this dead-end can be bypassed if the immortals progress by individual evolution. For example, they will have the means to change their own genetic material and their products. Already today, gene therapy can cure illnesses, creating a genetically improved individual. Similarly, immortals can add to their genome and proteome. Evolution can then continue in a static population of individuals.

But if an individual keeps changing and transforming for eons, assuming entirely new forms, then does the original individual still exist? In other words, did the individual actually achieve immortality? In the same vein, does the newborn exist in the eighty-year old man that he becomes? The definition of immortality is tied to the definition of the self. The immortals will have to define the nature of immortality that they desire.

As to Life itself, a population of permanent but static individuals will satisfy the purpose of Life, self-perpetuation. Since these static individuals will survive for eons, they will have also satisfied the Logic of Life, selection by survival.

Present humans cannot predict the thoughts of these future beings. Will they be happy? Will they be satisfied with a permanent but static existence? Will they still seek further progress? Can we judge now the morality of immortals eons into the future? Should we try to prescribe the ethics of immortals? Even if we attempt to do so, will our influence have any impact in a trillion years?

The history of the universe to the present is but a minute fraction of the habitable future, and too short to predict the long-term behavior of nature. It may or may not be possible for future life to change the laws of nature in their favor and to survive forever.

Our predictions cannot be proven. If the liveable universe comes to an end, there will be no intelligent observers to prove any predictions of its infinity or finiteness. Does non-existence exist if there cannot be an observer to observe it? The principle of observational equivalence suggests that non-existence, that cannot be measured by any physical interactions, has no physical meaning. In other words, after the matter of the universe has dissipated, even the ultimate non-existence of Life will itself not exist. It is not sure whether this is a form of immortality.

Every generation moves Life ahead by one step. While we desire that Life will be infinite, every generation can only take the next step. We can steer a course so that the next generation will strive to expand Life in the universe. Our faith in the future of Life may then be self-fulfilling.

The Re-Birth of Venus

Chapter 6.3 Limits of Knowledge

In this chapter we ponder the future of Life, which depends on the future of the universe. Is the future predictable in principle, and if so, can humans predict it?

Our ability to predict the future is limited by our knowledge. There are several types of limitation to human knowledge.

(1) Intrinsic limits, caused by types physical limits on measurement and the exchange of information.
(2) Biological limits, caused by the finite capacity of biological brains.
(3) Computational limits, caused by the amounts of computation required to construct models.
(4) Practical limits, where unfeasible action is needed to obtain information.

We will examine these limits, including the philosophical question; can we know the limits of our knowledge?

Intrinsic Limits

Physics imposes limitations on the material world that in turn limit our knowledge of nature. In fact, paradoxically, some of the great accomplishments of science are statements of limits. Amongst these are the impossibility of perpetual motion machines, and the laws of increasing entropy and randomness (thermodynamics, statistical mechanics); the indistinguishability of mass and energy, and of gravity and acceleration, and the impossibility of faster than light travel (relativity); and the impossibility of infinitely accurate measurements of velocity/momentum and time/energy (uncertainty principle, quantum mechanics).

Resulting from these limits, we cannot communicate with objects that are receding from us faster than the speed of light, and are beyond the event horizon. This will eventually include all other galaxies, and eventually all particles, when each particle has become a separated universe. We also cannot communicate with objects in black holes, and with other universes with which we may have had contact before the Big Bang. We cannot therefore physically prove, for example, if different physics can exist in these other universes, or if life will develop in them. We cannot receive information about the future of life in galaxies that recede beyond the event horizon.

This history can be consistent with matter made of infinitely sequence of sub-particles in sub-particles that is infinitely divisible and compressible. These particles could have passed through an infinite sequence of epochs when they were ever more compressed in space and time. Each particle on each sub-level will eventually become a separated universe made of its sub-particles, and each universe was a particle of a universe in the preceding epoch. This model will be discussed below. Events in these sub-worlds may affect our macroscopic world. However, we cannot exchange information with particles many of magnitude smaller than the particles of our world. Barriers to communication amongst different scales of magnitude may pose another intrinsic limit to knowledge.

Beyond the physical world lies mathematics whose laws govern matter. However, no system of axioms and proofs, and therefore no system of mathematics that we can construct, can be complete. Godel proved his theorem through a self-referential theorem that leads to a paradox.

Can we prove the limits of our knowledge? To ascertain these limits, we would have to know what it is that we cannot know. When we reached these limits, we would have to state that "We know everything that can be known" (1). To prove this statement, we would have to prove that "everything that we don't know cannot be known". However, we cannot prove anything about items that cannot be known, including whether or not they can be known.

To prove that we know everything, we would have to prove that "there cannot not exist anything that we don't know" (2). However, it is impossible to prove the non-existence of something that we don't know. Therefore, it is impossible to prove statement (1), and we cannot be sure if we know everything. For these reasons, we cannot prove if there are limits to knowledge and if we reached them.

Moreover, knowing if we know everything is itself an important item of knowledge. If we cannot know this fact, then we cannot know everything.

Biological Limits

Human knowledge is encoded by states of our brains. The biological brain is maid of a large but finite number of neurons, which can have a finite number of states. The brain can perform only a finite number of computations, and with limited speed.

Evidently, the universe contains more components than a part of it, the human brain. Therefore, the brain cannot fully model the universe. In fact, it can model only very small portions of nature. For example, we are far from being able to model accurately even a three-particle system. Even the brightest individuals know only miniscule portions of all the available information, which itself only touches even the principles, let alone the details, of nature.

In particular, we are still far from understanding the workings of a simple cell, much less a multicellular organism, or an ecosystem. If we cannot model a cell, we certainly cannot model our own brains. We therefore cannot model how much our brains can know, and what are the limits of its knowledge.

Computational Limits

Finite computers have similar limits as finite brains. They have a finite number of components that can work at finite speeds and assume a finite number of states. These are just a small subset of the components and possible states of the universe. Therefore computers cannot model all of nature.

Computational limits affect our understanding of complex systems. For example, a glass of water consists on the order of 10^{23} molecules, but physics can calculate accurately only the motion of two interacting particles, and even a three-body problem is intractable. The detailed structure of even one small biomolecule of 100 atoms is still beyond computation, and living cells consist of tens of thousands of such molecules and of their well as their complex interactions. Nature manages complexity but does not need to calculate it. As Einstein said: "Nature does not integrate numerically".

There are mathematical theorems and procedures that require very large or infinite numbers of computations, beyond any finite computer. Even if all the particles in the universe were incorporated in a computer that will work until matter decays, it would have a finite range of computations. For example, it may not be able to realize all the immense number of permutations of its possible states, i.e., the universe may not be able to play out in finite time all the possible futures that it could assume. It could not model itself or predict its own future.

Can We Predict the Cosmological Future?

To predict the future, we would have to assume that the laws of nature remain constant. Yet this is also uncertain. We can obtain physical evidence only from 14 billion years since the Big Bang. This is a minute fraction of the physical future, say of the 10^{37} years while ordinary baryons will exist. We may have to observe the universe for much longer in order to get a better indication of its future. Assume that our descendants observe the universe for another 10^{20} (one hundred million trillion) years. Even this is a minute fraction of the future of 10^{37} years of baryonic matter, still far too short make projections. In human terms, predicting the future of the universe from its past would extrapolate much longer than predicting an entire life from the first breath of an infant.

We therefore cannot predict the future of the universe in broad outlines. Much less can we predict its detailed future? The Uncertainty Principle precludes us measuring and predicting the exact states of quantum particles. We cannot measure with infinite precision the position and momentum of a particle (or the energy change and time-span of an event), and therefore we cannot predict with certainty the future position of the quantum object. We can predict only the probabilities of quantum events, but not their timing and outcome.

Much less can we predict, even in principle, the detailed future of the myriad of particles that make up macroscopic objects. In practice, we cannot have information on each of these particles, and even if we did, no human brain or finite computer could process that information to calculate the future.

Quantum events, with their uncertainties, affect and in fact comprise our macroscopic world. Therefore we cannot predict the detailed course of the future.

Can We Predict the Future of Life, and Transform Nature to Our Advantage?

The future of Life is bound up with the future of the physical universe. We wonder if Life can last forever, or, like every known pattern in this universe, if Life itself is also passing. How well can we predict?

If the universe had been eternal in the past, it would be reasonable that it could be also eternal in the future. However, evidence shows that the universe started in a vast inflation and explosion about 14 billion years ago, and that it expands at an accelerating rate. Together with the dissipating universe, all structured matter down to each baryonic particle seems destined to dissipate. The substance of this universe started as infinitely compressed and hot radiation, and may end as infinitely dilute and cold radiation.

It also cannot be known whether the fundamental constants of Nature such as the gravitational constants, the strengths of electromagnetic forces, even the speed of light, can change. It is not known what caused these constants, whether they could have been different, whether they can in principle assume different values, and if so, whether forces inside the universe are subject to these laws, including human action, can change them.

It is known, however, that the laws and constants of physics are balanced finely to allow life. Let us assume that we need to change the laws of gravity to prevent the ultimate dissipation or collapse of the universe. Could we change gravity without affecting the electromagnetic forces that govern biomolecules? It seems that our existence depends on a fine balance of forces, and that we couldn't change the laws of physics even minutely and survive.

In any case, predictions of the future can be verified or falsified by scientific observation only by being there. A prediction that Life will last forever can never be falsified by a living being, and a prediction that life will end cannot be falsified in finite time by a finite being.

Where There is Uncertainty, There Is Hope

What we do know for certain is that the laws of nature, and cosmology up to this time, allow life to exist. We know that universe contains vast resources for expanding life. The fate of the universe will remain in question for eons, and predictions of the finiteness of Life cannot be proven or disproved by scientific observation. Therefore, a long future for Life will remain scientifically possible for eons.

Even if this universe is habitable only for a finite time, there are speculations of multitudes of other universes. One of these is that the universe splits into two alternative universes with each quantum event, which is consistent with quantum mechanics. There are also theories that bubble universes are created and evolve forever. The sections below present another speculation, of infinitely embedded universes, where each particle expands into a universe of its subparticles in infinite successions. Each of these theories is consistent with physics but cannot be proven or falsified by observation.

These theories raise the possibility that some form of organized cognitive life exists in other universes or that we may implant life into other universes. In fact, if there are infinite universes of finite components that may constitute life, then an infinite number of life-bearing universes would have to exist. If our future is finite, we may take comfort in that we may be one in an infinite number of universes inhabited by an infinite chain of Life.

At the end, we cannot prove by scientific observation whether Life is finite or infinite, in this universe and others. The ethical consequence is that given these uncertainties, we must give the future the benefit of the doubt. We must do so since our predictions may well become self-fulfilling.

Limits of Knowledge - a Summary

By accepted standards, scientific theories should be "falsifiable experimentally " (Karl Popper). Some theories may be in principle subject to testing and falsification and therefore qualify as a scientific theory (+). Other theories may be impossible to test and falsify in principle (-) or it may be impossible to test them in practice within the foreseeable future, say within a million years (O). The following are examples of such statements.

Statement	Rating	Comments
Process X contributed to the origins of life	-	We cannot observe the past.
Life originates easily	O	We don't know and cannot prove how Life originated, and we cannot calculate this probability.
Extraterrestrial life exists	O	Cannot be falsified by finite searches.
Extraterrestrial life does not exist	+	Can be falsified by a single observation.
Life exists by star X	O	Definitive exploration will take too long.
Life can last forever	-	Cannot be falsified by a living observer.
Life will end	-	Cannot be falsified by a finite observer.
Other universes exist and contain life	-	We cannot communicate with other universes.
The universe is deterministic	-	We cannot repeat history to see if it could have been different.
There are limits to human knowledge	-	Finite beings cannot know if they know everything.

At the end, we must be satisfied with the knowledge that we have obtained and that we can enrich further. Understanding the knowable universe is sufficient for formulating our ethics, since our decisions will affect us in this universe.

Nevertheless, if err because of our limitations, then we must err on the side of caution. If we cannot be sure whether there is life elsewhere, then we must assume that we alone can secure life, and we must do so. If we cannot foresee how long our civilization will last, then we must expand life promptly. If we cannot be sure whether Life can be extended forever, then we should assume that it can, as our predictions can be self-fulfilling. We must assure that our descendants will exist, so that they can seek eternity.

Chapter 6.4 Can We Control the Future?

Limits Imposed by Physics

The limits of knowledge limit our powers: we can influence willfully only that what we understand. Within these limits, we try to influence the future in our favor. But in fact, can we affect the future at all? To answer this, we must know if the universe is deterministic. If it is, the future follows inexorable from the past, we can have no willful effect, and we have no moral responsibility to try.

According to quantum mechanics, particles and events are intrinsically statistical. For example, an electron may be anywhere, and its position is not fixed until it interacts with another particle. The decay time of an excited particle or the emission of a photon from an atom can be predicted only statistically, but not specifically. The quantum future is determined only statistically, and a virtually infinite number of actual futures may develop from the present.

Is the universe deterministic? Could its course have unfolded differently in the past, and must it follow a pre-determined course in the future? To test this, we would need to go back in time and observe if the universe can follow a different course. We cannot do this and therefore we cannot prove or disprove if Nature is deterministic. If it is not, it remains possible to affect the future, and we must act accordingly.

There are other basic limitations. The laws of nature show that we cannot create matter and energy from nothing, we cannot create engines of perpetual motion, and we cannot stop the trend of the universe to more randomness. We cannot travel faster than light and therefore we cannot affect parts of the universe beyond our event horizon. As the universe expands, more and more of the universe will recede beyond these horizons, past our influence.

We don't know how the laws and constants of physics originated and whether we can alter them. We certainly cannot alter the logic that lies behind the laws of nature, as this logic is immutable and permanent.

Limits on Controlling Future Biology

In independent worlds, branches of humankind will evolve in unpredictable directions. Even if one generation can design the next generation, it will not be able to predict the will of the next generation, and how they will design the generation that follows. This process may be called non-linear evolution.

Some future branches of self-engineered humans will succeed and flourish; some will go astray and perish. We may design the next step of evolution but we cannot foresee and control the long-term results of our designs. In the end, survival will choose the species that succeed. Even if human acts chose the paths, the logic of Life will shape the future and determine the outcome.

In general, it appears that we can control or predict events in the short run. On the long run, the possible details of the future may be numerous and unpredictable, maybe not even fundamentally deterministic. However, it appears that the overall course of the future, such as the expansion of the universe and the future of matter, may be predictable and unavoidable. However, determinism is beyond experimental knowledge, as we cannot create or observe multiple futures to examine if multiple different futures are possible.

The above are only a few examples why we cannot comprehend, much less control the course of Life or the fullness of Nature. Nevertheless, we achieved impressive depths of understanding. We know how the present universe started, what it is made of, how biology works, and what is our unique position in the universe. Within the boundaries that Nature allows us, we can take the correct step toward the future.

Chapter 6.5 Infinite Levels of Matter and Embedded Universes

Can the Universe be Both Finite and Infinite?

Intuitively, we expect the universe to have an infinite number of constituents and an infinite past and future. However, cosmology suggests that the contents of the universe are finite and so is its past.

Intuitively, we may expect matter to be divisible infinitely into smaller and smaller parts. However, physics shows that matter is divided into finite particles.

Can the philosophical expectations be reconciled with science? We shall outline a model of infinite embedded levels of matter and infinite embedded epochs of time, that nevertheless allows finite particles and a finite universe as observed from our level in our epoch. We shall observe that this model is consistent with observed science, has a universal Copernican symmetry, and may provide a mechanism in conventional finite terms for not well understood "instantaneous" processes.

In this model matter is divisible into particles and sub-particles through infinite levels of magnitude, and therefore are also infinitely compressible. This in turn allows the universe to have progressed through infinite embedded past expansions, yet in a very short finite time. As we look back, ever faster and hotter epochs converge to, but never reach, a point in time and infinite density.

Symmetrically, every particle on every level of magnitude will become an expanding universe in a future epoch. Its sub-particles on the next smaller level will expand in space as does matter in our epoch, until each of these sub-particles in turn becomes an isolated expanding universe of its sub-particles on the next smaller level, and so on. This infinite sequence of ever more slow-paced and colder epochs will expand toward, but never reach, a state of infinite expanse and zero density.

An observer anywhere looks into an infinite sequence of smaller and larger levels of matter and of shorter and hotter past, and longer and colder future, epochs. Therefore any observer on any level in any epoch sees themselves in the center of the levels of magnitude and epochs of the universe. The model has therefore a universal Copernican symmetry.

Can there be conscious observers on other levels of matter and in other epochs? The model of infinite embedded levels and epochs suggests that we may be a part of an infinite chain of life.

As we shall see, the model may also help to explain some features of relativity and quantum mechanics in relatively conventional terms. For example, an infinite series of compressions of matter on increasing compressed time-scales on infinite sub-levels can provide a mechanism for the conversion of mass into energy. These features can also provide a mechanism for the instantaneous collapse of wave-functions and quantum entanglement, as we shall show below.

Further, nature shows both aspects of determinism on the macroscopic level and statistical events on the microscopic level. Our

138

model suggests that physics is deterministic with infinite variables, which is on the border-line between deterministic and statistical.

The model of embedded levels and epochs can reconcile features of infinity and finiteness, and of determinism and statistical events that we observe or expect in Nature. However, the model deals with levels of matter and epochs of time with which we may not be able to communicate. If the model can be tested, it will be science; if not, it will be philosophy. In either case, it is consistent with current science, consistent logically, and philosophically pleasing.

Some Hints by Nature

Science concerns objects with which we can exchange physical information. However, Nature hints also of objects beyond our reach. For example, there is matter in black holes and in galaxies that recede beyond our event horizon. We cannot exchange information with these objects but we assign them physical existence.

Similarly, there may be worlds on scales of magnitude too small or too large, or events too fast or too slow, which we cannot observe. Nevertheless, Nature hints of their existence.

One such hint is that our biological world seems to be in the center between scales of magnitude very small and very large, between epochs of the universe that are very brief and very long, and between states of matter that are very compressed and hot or very dilute and cold.

Similarly, our biological time-scale of seconds, days and years seems to be in the center of time-scales. This may not be a coincidence. We may appear to be in the center because there was an infinite sequence of ever more brief epochs in the past and ever longer epochs in the future.

In fact, cosmology shows that our present universe is expanding from an immensely compressed, hot and short-lived state toward in immensely dilute, cold and long-lasting state. These may only be the limiting states of our epoch in an infinite progression of past and future epochs.

Matter could have converged toward infinite density in the past if matter is infinitely compressible. Physics suggests that before the Big Bang all the matter of the stars and galaxies was compressed into a size smaller than an atom. It is only a small extrapolation that matter may be infinitely compressible.

This would be possible if matter is infinitely divisible. Every particle on any level of magnitude is made of sub-particles, which are made of sub-sub-particles and so on. The particles on every level of magnitude are separated by vacuum. If the matter on each level collapses and eliminates the separating vacuum, and if events are faster on each sub-level of matter, then matter can converge in finite time through an infinite sequence of ever shorter, denser and hotter past epochs towards a point in time when matter was infinitely dense and hot.

In the other direction, our universe may have been a particle of a super-universe in the preceding epoch, which may have been in turn a particle in a super-super-universe in the next preceding epoch, and so on. These universe/ particles may have interacted in the past but separated with the expansion of the universe. Similarly, every particle in our universe will separate beyond the event horizon of all other particles and become an isolated expanding universe of its constituent particles on the next smaller scale of magnitudes, and so on for infinite levels of magnitude and epochs.

Can there be life on all levels of magnitude, and in each epoch? All epochs may contain moderate periods where particles of the level that constitute that epoch form self-propagating structures. It is pleasing to consider that we may be then part of an infinite chain of life.

The Main Features of the Model

- Matter is divisible infinitely into smaller and smaller components that exist on smaller and smaller levels of magnitude.
- On each level, matter is composed of particles separated by vacuum. Matter may be therefore compressed on each level. An infinite series of compressions on infinite levels would make matter converge toward a state of zero volume and infinite density and temperature.
- The pace of events may be faster on each smaller level, and an infinite series of compressions can occur in a finite time that is given by the sum of a convergent geometrical series.
- Compression releases energy, i.e., converts some of the apparent rest-mass on each level into energy. An infinite series of compressions on infinite embedded levels can eliminate matter by compressing it into zero volume, and releasing all the rest mass on all levels as energy. This may provide a mechanism for the relativistic conversion of mass into energy.
- Looking back into the past, infinitely divisible and compressible matter allows an infinite sequence of past epochs where matter was increasingly more dense and hot.
- The present epoch originated with matter in a state of immense density and temperature that underwent a short period of inflation

and explosion. This period may have been preceded by an infinite series of ever shorter, more dense and hotter epochs. However, the total duration of the infinite past epochs can be finite and very short, given by the sum of a convergent geometrical series. This view of an infinite number of ever shorter past epochs is seen looking back from any epoch.

- In the other direction, our epoch will be followed by a sequence of ever longer, more dilute and cold epochs in which the pace of events will be progressively slower. The duration of all future epochs will be an infinite divergent geometrical series that will expand toward, but never reach, a state of infinite cold and dilution. This view of infinite ever longer future epochs is seen when looking to the future from any epoch.

- A particle on any level of magnitude is a universe of its sub-components. When particles of one level in one epoch separate beyond their event horizons, the expansion of their sub-particles begins the new epoch.

- Similarly, each universe of a given epoch may have been a particle in a super-universe in the next larger level of magnitudes in the preceding epoch.

- This model provides a universal Copernican symmetry. The past is seen as an infinite sequence of ever shorter and hot epochs, and the future is seen as an infinte sequence of ever longer and colder epochs as viewed from any epoch. Matter consists of an infinite series of particles in particles on smaller levels, and of universes of universes on larger levels, as viewed from any level of magnitude.

- An entity on a given level of magnitude can interact only with other particles on the same level of magnitude. It cannot interact directly with sub-particles or with universes on other levels of magnitude. However, assemblies of particles on one level may affect events on the next higher level.

- Nature is deterministic, but with infinite variables. An infinite intelligence with infinite information about the states of all sub-particles on all levels could predict the future. However, a finite observer with information only of his own level will see the effects of smaller sub-levels as statistical. Determinism with infinite variables is on the borderline between determinism and statistical randomness.

- In any epoch and on any level of magnitude, forces among particles may form patterns of self-perpetuating "life". Maybe the inhabitants of one epoch can configure matter such that the next epoch will also allow "life". In this manner, we may be part of an infinite chain of life.

A more detailed and quantitative account of this model of infinite levels of matter and infinite embedded epochs of time is given in section A3.10 in the Appendix.

*N*ature hints that there may exist infinite successive epochs of the universe and infinite levels of embedded universes. This presents universal symmetry, where each observer in each epoch sees infinite, ever more compressed past epochs, and infinite, ever more expanding future epochs; and matter infinitely divisible and compressible matter of particle/universes on infinite levels of smaller and larger magnitude. Each epoch and each level contains large numbers of interacting particles that may allow structured "life". However, we cannot communicate with these preceding and succeeding, smaller and larger universes. Even in our universe, we may need to observe Nature for trillions of years to predict the future.

We have, nevertheless, gained much knowledge. We know that life is unique and precious. We are part of the family of complex organic life that can advance immensely in the future. Our distant descendants may find ways to extend Life ever further. As for us, we can take the next steps to assure that our descendants will be there to enjoy the future and to seek eternity. In our descendants, our existence will fulfil a cosmic purpose.

7

ASTROETHICS

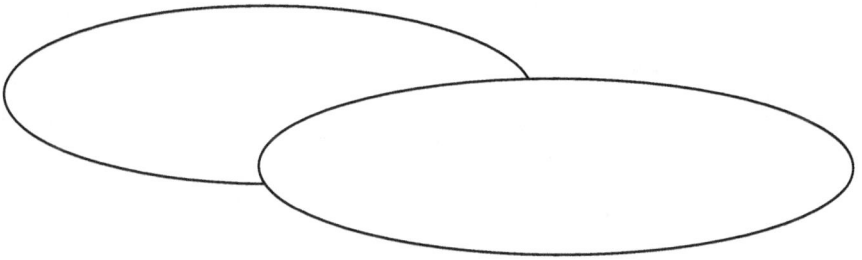

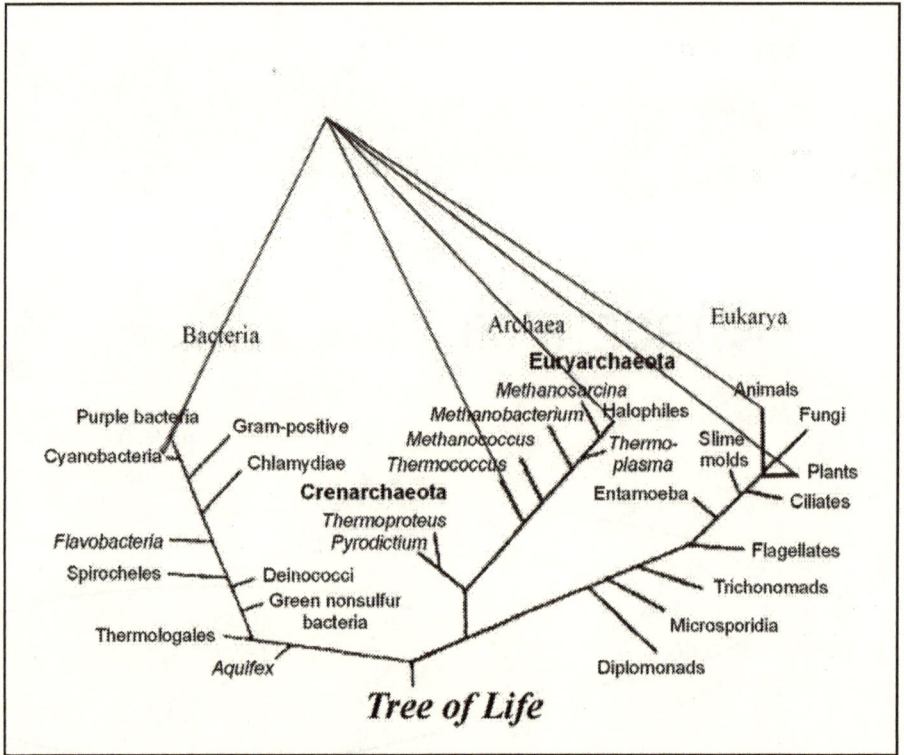

Bacteria

Purple bacteria
Cyanobacteria
Gram-positive
Chlamydiae
Flavobacteria
Spirocheles
Thermologales
Deinococci
Green nonsulfur bacteria
Aquifex

Crenarchaeota
Thermoproteus
Pyrodictium

Archaea

Euryarchaeota
Methanosarcina
Methanobacterium Halophiles
Methanococcus
Thermococcus
Thermo-plasma
Entamoeba

Eukarya

Animals
Fungi
Slime molds
Plants
Ciliates
Flagellates
Trichonomads
Microsporidia
Diplomonads

Tree of Life

As Life develops, our ethics will reflect the common roots and shared future of al life. This panbiotic ethics will encompass all life present and future, and strive to assure that life will pervade the universe.

Panbiotic Principles

and

Life-Centered Astroethics

*L*ife is a self-perpetuating process wrought in patterns of organic matter. Intrinsic to Life therefore is a purpose, self-propagation. Where there is Life, there is purpose.

The human purpose derives from the purpose of Life. It is the human purpose to propagate life and to assure that it will pervade the universe.

This panbiotic purpose implies the principles of ethics. That which promotes Life is good, and that which destroys Life is evil. The panbiotic ethics encompass all Life, present and future.

Pursuing the human purpose requires a coherent society. Panbiotic ethics supports the principles that cement society: love Life; respect reality; honor justice and human dignity.

In a self-designed future, our perceived purpose can become our destiny. Our ethics can become a force that will shape the human future, and with it the future of all life. Our actions, guided by our ethics, can even affect the physical future of the universe.

Ultimately, life can create the highest level of structure, a harmonious society. With life-centered ethics we can be a force of nature, to assure that life will achieve its full potential in the universe.

I. Nature

1.   The logic of Nature is immutable and permanent.
2.   The universe emerged from a point in space and time and contained compressed hot matter and energy. In the primordial universe, there existed the Laws of Nature.
3.   The universe exists in space-time and contains mass-energy and particles and fields.
4.   The basic particles are quarks and leptons. Gravity, electromagnetism, and weak and strong nuclear forces act amongst the particles.
5.   All of the universe that is accessible to human experience is made of these particles and forces, of the patterns that they form, and of the laws of Nature that reside within them.

II. Life and Purpose

6.   Life is a process wrought in organic matter that self-propagates through gene/protein cycles.
7.   The processes of Life lead to survival and propagation. But action that leads to a selected outcome is equivalent to the pursuit of a purpose.
8.   Intrinsic to Life there is therefore purpose. The purpose of Life is to continue to live.
9.   A universe that contains Life, contains purpose.

III. The Human Purpose

10.   Humankind belongs to the family of Life. The human purpose is therefore one with Life's purpose.
11.   Therefore the human purpose is to forever propagate Life, and to elevate Life into a dominant force in Nature.

IV. The Logic of Life

12.   Those living patterns who propagate successfully survive; those who do not, vanish. This is the logic of Life.
13.   The logic of Life is self-evident, immutable and permanent.
14.   Self-reproduction creates new life-forms through random mutations. Those new life-forms who propagate best multiply, and survive longest. Thus evolution leads to progress.

15. Therefore the logic of Life defines progress as the increasing capacity for survival and propagation.
16. All life-forms are tested by survival, whatever means they arise by.
17. The same logic that controlled natural evolution will control engineered evolution.
18. The patterns of Life forever change, but the logic of Life is forever permanent.

V. The Family of Life

19. Who belongs to the family of Life? Only sentient humans can ask and answer this question. Human judgement defines fellow Life.
20. The processes of organic Life are enacted by proteins, whose structures are encoded by genetic sequences. The core of the living process is the creation, through DNA codes, of proteins that act to reproduce the DNA codes.
21. Our family of organic Life contains all the organisms who reproduce through gene/protein cycles.
22. Humans possess an innate empathy to other living beings. Human emotions, ethics, rational judgement and religion all value Life.
23. All the human faculties: rational judgement, emotions, moral codes and the religions that rest upon them, support ethics that promote our family of organic Life.

VI. The Value of Life

24. Value judgements are formed by intelligent beings. The value of Life must be self-defined by living beings.
25. To living beings, Life itself is the highest value. From the human perspective, Life has the highest value amongst all the contents of the universe.
26. Living systems are uniquely complex, and precisely tuned for their function. The harmony of this complex system endows Life with a unique value amongst all the creations of Nature.
27. Biological structure would be unstable if the constants and laws of Nature were even slightly different. Life can exist only by a precise coincidence of the basic laws of matter.
28. In this sense, the physical universe itself converges to a unique point in Life.
29. Life has a unique value both from its own vantage-point and in the overall constitution of the universe.

VII. Self-Design and Survival

30. Ethics must guide human action to fulfil the human purpose.
31. Those who strive best to survive persist, and reproduce most. The logic of selection installed thereby the instincts of survival in all species. These instincts are the foundations of human nature.
32. The instincts of survival may be enhanced or lost through self-engineered evolution.
33. Our descendants will survive only if the will to propagate Life is always consciously installed in human nature, from generation to generation.
34. Adaptation to space will require self-directed evolution, and space can accommodate the diverging products of self-engineered life. Self-directed evolution and expansion in space support each other and are mutually dependent.
35. Directed evolution will allow progress without pain. The pains of random evolution and extinction will be no longer needed.
36. In harsh space organic Life will integrate with robots, who must serve Life but not replace it. Living beings who would replace themselves by robots would have betrayed the purpose of Life, self-propagation.
37. When humans integrate with robots, control must remain in organic brains with vested interest in the survival of organic gene/protein life.
38. As humans and science advance, all that is allowed by Nature will be achieved. That human purpose which will be sought, will be accomplished. Our perceived human destiny will be self-fulfilling.
39. Self-governing species can survive only they seek survival deliberately, and must therefore supplement the instincts of survival with conscious biotic ethics.
40. The tests of survival are endless. Humans will always need to pursue deliberately the duty to Life inscribed in our every cell.
41. Therefore this principle must be the foundation of ethics passed on from generation to generation: that the human purpose is to forever propagate Life.

VIII. Good and Evil

42. From the human purpose derive the axioms of ethics: That which promotes life is good. That which destroys life is evil.
43. These values derive from a biotic ethics that encompasses life, present and future:
44. Nurture the young, since they are the seed of the future.
45. Assist Life where it falters: Help the needy, feed the hungry, cure the ill.
46. Advance the richness of living patterns. Conserve and enhance the diversity of living species and develop ever more advanced life forms.
47. Plant Life where it is absent.
48. Seed with life all space and time, and encompass in Life all the resources of the universe.

IX. Truth, Information and Survival

49. Structured living matter contains order, but the same matter could be arranged in many more ways that would be lifeless.
50. Therefore living structure is improbable. Life can exist just narrowly, and living patterns are improbable and fragile.
51. The ways of survival are therefore a select few, while many more false paths lead to death.
52. The paths of survival require information and selection on many levels:
53. To construct and maintain biological structure; find sustaining elements in a random environment; to chose the optimal mates amongst a chance population; to select the fit amongst random mutants.
54. The interplay of chance and selection is intrinsic to the living process.
55. Survival and progress demand purposeful action directed by true information.
56. True information is therefore vital.
57. Reality must be recognized and confronted by action. Self-deceit may comfort and lead smiling to death. Nature does not forgive ignorance.
58. Truth, information, and the knowledge of reality serve Life.

X. Survival and Peace

59. Where there is Life, there is hope. Good can always emerge after evil.
60. Survival must be won moment by moment, day by day, year by year, eon by eon; but extinction once alone is the ultimate defeat, and only death can win a final victory.
61. No passing cause can justify the risk of our extinction; as where there is no life, there is no hope, nor does justice have meaning.
62. Only humans can secure the future of Life. Human extinction would be total extinction.
63. Ultimate total war is ultimate evil. No passing cause, even the pursuit of justice, can justify this ultimate danger.
64. Life is fragile. A technological society who can expand in space will also have the power of self-destruction. Aggression, that served evolution in the past, can now threaten the future.
65. Space can accommodate many cultures and species. Territorial aggression will be obsolete.
66. Therefore the ethics of Life commends peace.
67. Life will need cooperation in space, and advanced societies cemented by equitable justice.
68. Peace, justice, dignity and freedom promote Life.

XI. The Value of an Individual

69. The value of a human life is measured by its contributions to the human purpose.
70. The chain of life is constructed of day-to-day survival, and every constructive human life that contributes to the chain, contributes to the human purpose.
71. Small events may decide the evolution of complex systems. Each individual life can affect the future.
72. The imprint of every human being will live on in the course of the future.
73. The imprint of a human being on the patterns of future is our share in the worlds to come.
74. The senses of pleasure and happiness evolved under the logic of natural selection.
75. Therefore those who find wholesome pleasure in health, sustenance, safety and procreation, and pursue these ends, will survive and propagate best:

76. But self-serving indulgence mislead from the works of survival, and happy delusions can lead smiling to death.
77. The pursuit of healthy pleasure and happiness serve life and are morally good. Denying healthy pleasures and happiness, to oneself or to others, is evil.
78. As well, a happy society treasures peace and equitable sharing.
79. Life-centered ethics commends the pursuit of happiness.

XII.   Natural Justice and Human Justice

80. Natural law recognizes cause and effect only.
81. But human justice seeks fairness and compassion, rewards good and punishes evil.
82. Natural law judges an act by its effects and it is blind to intent.
83. But human justice weights both the effect and the intent.
84. The justice of Nature does not forgive.
85. But human justice forgives, and honors repentance.
86. Natural justice is collective: an efficient society will flourish, and individuals will prosper regardless of merit.
87. But human justice awards individual justice; it rewards or punishes by individual merit.
88. The laws of Nature can be used but not defied and human justice can exist only in the framework of Nature.
89. Human justice will increase with human power, and moral justice will rule the future.

XIII.   Free Human Will

90. In the quantum world many futures can derive from the present and if multiple paths can unfold, human action may affect the future.
91. Human impulses are neural events evoked by internal processes and external information. Our impulses appear to us as our will, and when exercised freely, the outcome is judged by survival.
92. Thereby freedom supports progress and Life benefits from freedom.
93. The empires of Life will be ruled by freedom, truth, happiness, human justice, and peace.

XIV. The Future of Life

94. Life that we plant in new worlds will fill the universe.
95. The paths of Life will diverge in multiple worlds, to explore the limitless scope of structures that Nature permits.
96. The family of organic Life will grow in quantity, complexity and diversity.
97. Our family of organic life will be secure as our seed fills the galaxy.
98. Life, born of the laws of Nature, will permeate the universe.
99. Our descendants will fill all habitable space and time and seek to extend Life to eternity.
100. When Life permeates the universe, our human existence will find a cosmic purpose.

8

THE PANBIOTIC PRINCIPLE

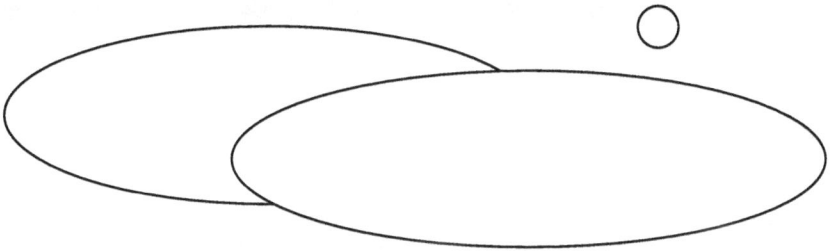

The Re-Birth of Venus symbolises the emergence of self-directed conscious life in space. The success of self-designed conscious life will depend on its guiding ethics. Principles focussed on the survival of life will secure survival.

The following is a poetic presentation of life-centered ethics. *(Published as "Principles of Life and Survival", Precept Press, Maryland, 1984).* A systematic description of life-centered ethics was given in the preceding chapter.

"*F*orever safeguard and propagate life; establish life throughout the universe."

"The fate of Life is entwined with human destiny."

"The ways of survival are few, and the love of Life will help to find them: Preserve the love of life in the foundations of human nature."

"Safeguard Life, and our seed will inherit the future."

"The fit survives and inherits the future; the unfit perish. This is the Logic of Life; it is a law of Nature, immutable, and forever permanent."

"Life is the matrix of all love, justice, beauty, and wisdom."

"In a self-designed future, our expectations will be self-fulfilling. Where there is life, there is hope; and where there is hope, there will be life."

"Plant the seeds of life throughout Nature and your seed will persist in the Land of Existence forever."

"Three principles of wisdom: love life; respect reality; and honor human dignity."

"The universe will satisfy all, and no living being will need to harm another living being: Justice, love and peace will prevail."

"All space and time, energy and matter will be part of Life or serve its survival."

"Our seed will conquer the land of existence forever; only remember to instill the love of Life in each generation. Always follow the commandments of Life that are inherent in every living being."

1. The Human Purpose

1. A message of Life to future generations.
2. Human being, this is your purpose: to forever safeguard and propagate life.
3. As the patterns of life actively seek continuation in the future.
4. The purpose of life is perpetual survival and propagation. Therefore, where there is life, there is purpose.
5. Therefore this commandment is inherent in every living being:
6. Forever safeguard and propagate life; establish the dominance of life over all nature; conquer for life the universe.

2. Propagate the Will of Life

1. The fate of Life is intertwined with human destiny.
2. Human power and human will must secure life's survival.
3. Perpetuate therefore and follow the will of life within your living being.
4. Inscribe the Commandment of Life as the driving will of each following generation;
5. Instill this commandment foremost it foremost in the mind of your descendants,
6. Teach them to those descendants whom you beget, and those whom you will design to follow.
7. Make this commandment the principal of your every deed and the purpose of your every thought.
8. Inscribe the Commandments of Life as the basic precepts of your laws and establish the love of life as the first tenet of your ethics.

9. Always remember that the ways of survival are few, and only the love of life within you will find them.

10. Use with the awe of life and death in your heart the power to create new humans:

11. Preserve the love of life and the fear of death: as the instincts of life are the guardians of survival.

12. Preserve therefore foremost in the foundation of human nature the will to propagate and perpetuate life.

3.   Beware the Numerous Ways of Extinction

1. Follow the Commandment of Life and your seed will inherit the future.

2. The treasures of the universe are within your reach.

3. Beware therefore lest you allow the will of life to wane within you from generation to generation.

4. Lest your desert the commandments of life by the lack of will or by ignorance;

5. Or lest you threaten life through destructive arrogance and hatred;

6. And lest you endanger an eternal future by a blinding greed for passing pleasures and power.

7. As the decay of death is more probable than the order of life; the ways of death are far more numerous than the ways of survival.

8. Remember that life wins only when it continues;

9. Life must win day by day, minute by minute; only death can have a final victory.

10. Therefore without the wisdom of life you will lose your way and perish.

11. You will destroy your own kind in blind aggression or wither in lethargy;

12. You will perish confused amongst the many open and hidden paths that lead to death.

13. The fit will survive and inherit the future; the unfit perish. This is the Logic of Life,

14. It is a law of Nature, immutable, inescapable and forever permanent.

15. Chose Life and you will live; abandon the will to survive, and by the Logic of Life, you will perish.

16. Beware therefore lest you allow the love of life to grow weak within you from generation to generation.

4. The Empires of Life Will Encompass the Universe

1. From our perspective, human existence is the center and purpose of the universe.

2. Therefore fill the cosmos with life: multiply and populate the firmaments.

3. Preserve your inherent desire for the land; as it will plant your seed in far dominions.

4. The rich resources of space will be transformed in your hands into cells, tissues and organs, bone, muscle and brain, the substance of life; and into energy and matter, metals, crystals and machines to serve you.

5. Secured in many worlds, your future can never be destroyed.

6. Secure your seed in the many empires of life, and the dread of extinction will not be remembered.

7. Throughout the universe, many new worlds will rise to implement the human ideals of truth, peace and justice.

8. In the expanses of space, no person will need to vanquish another person; no nation will need to subdue another nation, no species will need to extinguish another species to survive;

9. Diverse species can arise from your seed and coexist securely.

10. Peace will prevail throughout the heavens.

11. Your seed will rule all time and space, and construct new human dimensions of beauty and knowledge.

12. Throughout space will rise empires of freedom

13. Humans will establish the dominance of life throughout the universe.

14. Therefore bring Life into the expectant but empty space,

15. Build the empires of life and expand the highest orders of structure: societies of conscious beings, living in truth, peace, compassion and justice.

5. The Value of a Human Life

1. Human existence is the highest level of being; therefore treasure life.

2. Human survival is our dignity

3. Each human life is as precious as the eternal laws of Nature; as even the laws of Nature acquire conscious existence when reflected in human knowledge.

4. Hereafter, human future and Life are intertwined. The future will take shape through human power, or life will perish if deserted by humans;

5. Therefore human life is as dear as the future of all life.

6. Therefore relate to all humans with respect, compassion and justice;

7. As what bonds you to life, beyond a few hours of pleasure and beauty; but the love of life and the closeness of other humans.

8. Therefore if you wrong others, repair the damage and never repeat that evil.

9. Always adhere to truth;

10. As trust cements social structure which will implement the future.

11. Always make human love the source of your deepest satisfaction;

12. Take pleasure in the pursuit of happiness.

13. Pursue justice, compassion, peace, truth and the love of life; and you will find a lasting future.

14. By the Logic of Life, you find the ultimate joy in raising the young; as the desire to procreate promotes survival from generation from generation.

15. Therefore it must always be your conscious joy to give new life and to protect and raise the young. The Logic of Life will then assure that your seed will forever continue to prosper.

6. The Purpose of Life is Self-Defined

1. Search the meaning of your life within your living self.

2. You will not find it in the inanimate universe

3. You yourself must define your purpose; human faculties must make the judgement.

4. Ask the love of Life our instincts, emotions and intellect to define our purpose.

5. Follow Life's commandment; you will always find it in the foundations of your being.

6. Beware lest abstract doctrines replace the Commandment of Life.

7. Expand and perpetuate the Family of Life;

8. As the values that bind together create the highest structures, and they exist in living beings.

9. Life is the matrix of all love, justice, beauty and wisdom.

7. Confront Every Threat

1. Weigh your every deed by its value for human survival.

2. Uphold and pursue every enterprise that promotes human survival and progress

3. Do not hesitate to destroy every threat.

4. Never risk the infinite promise of the future for short-lived powers or pleasures.

5. If a threat arises to imperil all life

6. From the external forces of Nature, or through human hands; even a remote danger

7. Never hesitate to rise up and destroy such a threat:

8. As the suffering of a living being is bad; and the death of a human being is evil

9. Then surely a threat to all life is an abomination that cannot be tolerated under the heavens.

10. Therefore if you see a threat to all life

11. Forged in arrogance, in ignorance or in evil;

12. Forged by an individual, a group, evil thousands, or mighty nations;

13. Oppose them and do not remain silent.

14. As silence is an evil that will turn upon you in vengeance.

15. Confront the approaching dangers; do not close your eyes, do not ignore the future to sooth the present;

16. Always beware of self-deceit, as it will lead you smiling to death.

8. The Legacy of a Human Being

1. Expect hope from your own wisdom, as you alone possess the powers to assure Life's survival

2. Fear only your own failure, as only self-extinction can deprive you of your share in the future.

3. Always hope for Life and reject despair:

4. Remember that both hope and despair can turn into self-fulfilling prophecy.

5. Beware of projections of doom that predict that your universe will collapse or vanish; or that claim to know that human existence must come to an end.

6. The truths of the universe are deep and the ultimate future is remote, and you cannot know it for certain.

7. Where there is Life there is hope, and where there is hope there will be Life;

8. Your descendants will find unforeseen powers to make this cosmos the lasting matrix of life's patterns.

9. With the love of your warm womb remember the Commandment of Life.

10. Since you know as surely as your soul is entwined with the soul of your children: that if Life should ever end, your life has been wasted.

11. Therefore if our species will ever vanish, already now we exist in vain.

12. As the value of each human life is its lasting effects:

13. You live on in your descendants and in the imprint of your life on the patterns of the future.

14. The impact of each human life will multiply as Life grows

15. Your imprints on the patterns of the future are your share in the worlds to come.

9. Life is Good, Death is Evil

1. The order of life is the essence of good; the forces of chaos and extinction are evil.

2. The instincts of life will move you to protect the young.

3. Compassion moves us to cure the sick rather than witness his agony;

4. To help the poor rather than to ignore his hunger;

5. To love rather than to hate;

6. To build rather than to ruin;

7. To comfort rather than torture;

8. To help a life soar rather than stand idly by and see it falter.

9. Every battle between good and evil is the struggle of the forces of life against the forces of destruction.

10. Beware, as both good and evil inhabit human nature.

11. Many human ways can be moral if they promote life.

12. Many diverse pleasures, dreams, customs and faiths may be moral, as survival and progress may be sustained in diverse human frameworks.

13. Every human life is spent well if it helps life to populate the future.

14. Therefore this is the moral instruction of the Commandment of Life:

15. Uphold your duty to the forces which brought you into being; to the patterns of life imprinted in your self; to the Family of Life whose fate is in your hands.

16. Always promote life, as this is the essence of all good; and oppose death, which is the essence of evil.

17. Establish the dominance of life over all nature

18. Plant the seeds of life throughout the universe, and persist in the Land of Existence forever.

10. Future Humans Must Value Survival

1. Now soon times will come and the secrets of life will all open before you

2. Then the complex mysteries of your existence will open before your eyes;

3. In those days your hands will carve out cells and tissues, new flesh and blood and brain, descendants whose likenesses have not lived before.

4. But in those days remember that the breath of life is as dear when you perceive it as complex molecular processes, as it is dear now in the breath of your infants.

5. The glow of life that transpires from intricate molecular patterns is the same glow that reflects in the hope on your children's faces.

6. When Life will advance by benign design, progress will cost not the pain of Nature's unsuccessful errors.

7. In your wisdom, create new branches of life that will populate the diverse worlds of the universe:

8. Life must not depend upon a single species for survival.

9. Do not hesitate therefore to constitute new human traits, organs, tissues, genes and minds that will flourish in many worlds,

10. As the human spirit can be embodied in many diverse living frameworks.

11. Only always beware to imprint in your descendants these attributes, as in them lies the unique human power:

12. Aggression to conquer new worlds, rational intellect, and compassion in proper balance - and forever, the love of life.

11. Pursue Peace

1. The dreaded peril of an empty universe resides in a species who possess both power and hatred.

2. When envy, fear and mistrust rule, to the voice of life itself you may then turn deaf ears;

3. You hasten to ruin; your ears are rocks to the voice of life within your heart.

4. You say: Only my own peace is just; only my own faith is the truth; only my own wealth is deserved; only my own dignity must be honored; only my own pride must be revered.

5. For my own justice I will die; for my own peace I will fight;

6. For my own faith I will kill; for my own wealth I will devastate; for my own dignity I will humble others; for my own freedom I will oppress.

7. But is there justice in death; is there peace in waste;

8. Will killing inspire faith; is there gain in destruction; is there pride in terror;

9. Where there is death, who is free?

10. No; death is evil: its path is the ugly shame of an empty universe.

12. Respect Reality

1. Respect reality

2. Recognize and eliminate your self-destructive powers.

3. Do not deny self-created dangers, as self-deceit can compound the destructive powers of hatred.

4. When the threat of death hangs above your homes, do not take comfort in denial and silence

5. Open your eyes when destruction stands erect on your doorstop; when self-destruction looms do not choose to dream.

6. Where there are weapons that threaten all Life for short-lived ideals, do not look away and remain silent.

7. If your heart is blind in complacent folly, your ears are deaf and your voice is frozen; if you forfeit the voice of life within you:

8. Then you can destroy eons of hope for short-lived greed and transient ideals;

9. Then by your own hands death may come therefore from the skies

10. Your sacred homes may then crumble

11. Death will glow in the blind cities; and rot in abandoned fields

12. Your tender young will lay in waste amidst your ashen hopes

13. The clouds will rain pain, the faint spring breeze will bring vanished dreams

14. Your wisdom and your dreams, your hatred,

15. Your brave hearts, your fearless self-deceit, your great promise, your lies, your blind loves, your hopes will all vanish.

16. Listen therefore the voice of life within you; when it says cry out, do not remain idle.

17. Therefore if a threat imperils all life, eradicate that danger; do not remain silent:

18. As silence is an evil that can turn upon you and on your children in vengeance.

19. Always rise up to destroy every threat to Life.

13. Preserve the Love of Life

1. Now the times will soon come and the brain and soul of the next generations will be shaped by your own designs.

2. Beware then lest you design descendants lost in the worship of pleasures; lest the healthy pleasures that binds you to life turn self-serving.

3. Lest your descendants submerge in the beauties of their soul, the lofty cathedrals of abstract theories and elegant equations, and forget what is real;

4. Lest they worship pleasures and forget the future.

5. There is also death in the comfort of lethargic fatalism; if you leave your fate to chance or resign to false notions of a predetermined future;

6. Beware lest you choose to believe that your fate is sealed and not subject to your will.

7. Beware lest you believe that you lack a higher purpose.

8. Do not build docile flesh-machines of happiness who will never feel the agonies of ambition, yearning and desire.

9. When you create intelligent robots to serve you: creatures not of living tissue, beings who are durable and intelligent but alien;

10. Your own creations may turn against you to replace our family of organic life.

11. Make sure that the ultimate control remains in organic brains with self-interest in biological survival.

12. Beware lest your engineer your own end and betray the force of life who brought you into being.

13. Do not lose challenge when the secrets of nature have been reduced to formulas and all beauty has been long explored;

14. Lest your descendants will be lost in a dead-ended paradise, in a perfect world that is aimless.

15. Beware also of self-proclaimed Messiahs who may risk all Life to serve their own ideals;

16. Blind prophets who do not see that where there is no life there is no justice, no good, no evil, no love, no hatred, no righteousness, no sin, no sorrow, no joy; only indifferent void.

17. As if you yield to death, wherefore were your dignity, your hope, your pains, your promise?

18. Therefore beware when the time comes to design the body, brain and mind of your descendants;

19. Make sure that the instincts of survival should never atrophy.

20. Always preserve in the foundations of human nature the commandment to safeguard and propagate Life.

14. Forever Safeguard and Propage Life

1. The future of Life resides in human wisdom.

2. Soon days will come and you will understand that life is good and death is evil.

3. You will know that peace and compassion lead to survival.

4. By these three Commandments of Wisdom will you prevail: Love Life; Respect Reality; and Honor Human Dignity.

5. The Commandment of Life must be your lasting faith.

6. You will wish one wealth: the joy of existence, and one dignity: the pride of survival.

7. In those days you will aspire to the warmth of stars and reach them

8. Tender infants and Life's sages will in the light of new suns flourish.

9. As life is safe in numbers; as the trees send up their many seeds in the wind and as the fish send their many seeds in the currents;

10. So will your seed multiply amongst the stars to assure survival against extinction.

11. The good in humans, and not the evil, will continue in your descendants in many and diverse worlds.

12. In those days the treasures of the universe will satisfy all, and no living creature will need to destroy another living creature.

13. The human harmony of justice, love and peace will rule the firmaments.

14. All space and time, energy and matter will be part of Life or serve the survival of Life.

15. New knowledge will breed new powers to perpetuate this physical universe and the precious life that it contains.

16. Your seed will conquer the land of existence forever.

17. Only always remember to instill the love of Life in the conscious will of each rising generation;

18. Preserve in the foundation of your morals, laws and conscience the commandments of Life:

19. To forever safeguard and perpetuate Life; to establish the dominance of Life over all nature; to conquer for Life the universe; and to seek for Life eternity.

20. Forever perpetuate the commandments of Life that are inscribed in your living being.

GLOSSARY

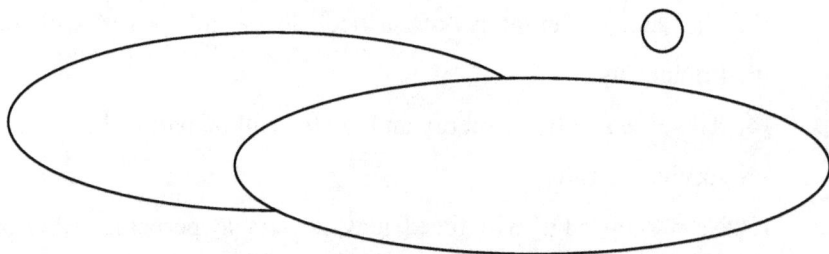

Glossary	
Accretion Disks	Disks of accreting gas and dust about new stars.
Aerobic microorganisms	Microorganisms that require oxygen
Albedo	The fraction of incident light that an object reflects.
Amphiphile	A molecule that can dissolve both in water and fats.
Anaerobic	Microorganisms that *do not* require oxygen
Areal Density	The mass of a unit area of a film or layer of material.
Astroecology	The scientific study of interactions between biota and space environments
AU	Astronomical Unit. The distance from the Sun to the Earth, $1.5x10^{11}$ meters
Autotrophs	Microorganisms that manufacture their own organics, such as photosynthetic bacteria and algae
Baryonic matter	Ordinary matter, protons and neutrons that make up the nuclei of atoms
Bioavailable materials	Materials that are available to taken up as nutrients, such as to plant roots, or that can be made available by natural processes such as leaching
Biomass	The mass of biological organisms
Biota	Life forms, including both animals and plants
BIOTA	Biomass Integrated Over Times Available. A measure of the total biomass that will have lived during the habitable timespan of an ecosystem
Carbonaceous meteorites	Several types of meteorites that contain carbon-based organic molecules. The CI, CM2 and CV3 types contain about 10, 2 and 0.2 percents organics by weight
Colony-forming units (CFUs)	Live microorganisms can form colonies on agar counting plates. CFU units are used to count the number of viable microorganisms per ☐illilitre of solution (CFU/ml)
Cometary nuclei	Objects made of rock, ice and some organics, usually about 10 km across, located mostly in the Kuiper Belt and Oor Cloud. Some leave these belts for orbits in which they pass near the Sun, where materials vaporize from their surfaces and they become comets with tails

171

Glossary	
Cryconite dust	Interplanetary dust particles that landed on Earth
Cyanobacteria	Blue-green algae, primitive cells without nuclei, that have existed on earth for 3.5 billion years
Cyborgs	Composites of human and robot parts
Dark Cloud Fragments	Fragments of an interstellar clouds with relatively dense material.
Dyson Spheres	Spheres proposed by Freeman Dyson that advanced civilisations can construct about atars to capture all of their radiation
Ecoforming	Formation of new ecosystems
Gene/protein life	Biological life-forms that reproduce using a DNA sequences in genes that code for proteins which directly or indirectly help to reproduce the DNA code
Gigayears, GY, eons	A period of one billion years
Heterotrophic bacteria and fungi	Organisms that require organic materials for food, cannot produce their own
K Kelvin	Units of temperature. Degrees Kelvin are similar to Centigrades but start at absolute zero at -273 C.
Kg-years	Biomass (in kilograms) that survives for a certain number of years. The product of mass times years gives a measure of the total amount of life that have existed during a given period of time.
Kuiper Belt	A belt of comets beyond the orbit of Pluto but closer than the Oort cloud.
ly Light Year	The distance light travels in one earth year.
Macronutrients	The main nutrients for plants, e.g., calcium, magnesium, potassium, nitrate, phosphate, sulfate
Meteorite extracts	Solutions obtained by extracting meteorites in water or other solvents
Microcosms	Small biospheres that model larger ecosystems

Glossary	
Micronutrients	Minor plant nutrients such as iron, manganese
Oort Cloud	A cloud of comets orbiting the Sun at a large distance. The Oort cloud is the source of many observed comets.
Organics	Molecules that contain carbon atoms, not necessarily of biological origin
Panbiotic program and ethics	A program or principles that aim to establish life through all habitable environments in the universe
Panspermia and directed panspermia	The spread of biological organisms, especially microorganisms in space by natural processes or by designed missions
Planetary materials	Materials found on major planets and on minor planets such as asteroids and comets. Usually refers to planets other than Earth
Protostellar condensation	Star-forming areas in dense parts of interstellar clouds
Self-sufficient Humans	Humans who can create/supply their own energy or nutrient needs through such processes as photosynthesis
SIU units	The International Standard Units using a meteric system (kilograms, meters, seconds, Watts)
Solar Sails	Thin films stretched out in space that are propelled by solar radiation
Solys	Shorthand for solar system. May be used as a term specifically for inhabited solar systems
Swarm Strategy	A strategy to seed other solar system by directed panspermia using a large swarm of small microbial capsules
Terraforming	The conversion of planets and asteroids to form habitable Earth-like environments
Thermophiles	Microorganisms that live in hot environments
Time-integrated biomass	Biomass (in kilograms) that survives for a certain number of years. The product of mass times years gives a measure of the total amount of life that have existed during a given period of time.

APPENDIX:
TECHNICAL PAPERS

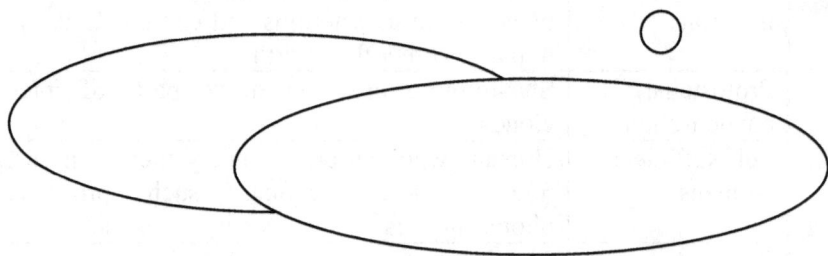

Appendix: Table of Contents

Appendix 1. Astroecology: Tables

Table A.1 Terms and data relevant to astroecology

Resources	Units	Quantity	Definitions and formulas
Carbonaceous Asteroids	Kg	10^{22}	
Comets	Kg	10^{24}	
Limiting Nutrient			The nutrient that allows the least amount of biomass
Nutrient Concentrations And Elemental Contents	g/kg	Table A2.2	The content (grams) of a nutrient element per kg of a resource material
Usage Or Waste	kg/year	$k_{waste} M_{biomass}$	Amount of biomass used or lost as waste, per year
Baryonic Matter	Kg	10^{41} (galaxy), 10^{52} (universe)	
Biomass			
Biomass Constructed From Asteroids/ Comets	g/kg	Table A2.4: Average biomass, 0.48 g/kg from extractable materials, 60 g/kg from total elemental contents; Human biomass, 0.25 g/kg from extractable materials, 31 g/kg from total	The amount of biomass (grams) that can be constructed from 1 kg of resource materials according to element x

Resources	Units	Quantity	Definitions and formulas
		contents (based on N as limiting nutrient)	
Biomass Of Human Individual	kg/individual	50	
Supporting Biota	kg/individual	0 (independent humans); 1,000 (supported humans)	Biomass required to support a human individual.
Time-Integrated Biomass (Biota$_{int}$)	kg-years	Table A2.5	$$BIOTA_{int} = \int_{t_o}^{t_f} M_{biomass,\,t}\,dt$$
Time-Integrated Populations	human-years	Table A2.5	
Elemental Contents In Biomass	g/kg	Table A2.3	
Baryonic Matter	Kg	10^{41} (galaxy), 10^{52} (universe)	

Power Sources And Requirements			

Resources	Units	Quantity	Definitions and formulas
Luminosity	Watts		The power output of a star, i.e., energy output per second
Luminosity - Sun	Watts	3.8×10^{26}	Mass: 10^{30} kg Life-time: 5×10^9 years
Luminosity - red dwarf	Watts		
Luminosity - White Dwarf	Watts	10^{15}	Life-time: 10^{20} years
Power Requirement Of Biomass	Watts/kg	2 (human), 200 (bacteria), 100 used here;	Power from energy source, including conversion efficiency,

Resources	Units	Quantity	Definitions and formulas
		or 1,000/per person (industrial society)	required by one kg of biomass
Ethics			
Biocentric Ethics			Life-centered ethics
Biotic Ethics			Ethics that encompasses all life present and future, and seeks its expansion
Panbiotic Ethics			Ethics that seek to maximize the amount of life and to include all the resources in the universe in living matter.

Table A.2 Contents of nutrient elements, organic carbon, and water in meteorites and in some terrestrial materials (gram/kilogram).

Extractable Contents[a]	C	N	S	P	Ca	Mg	K	Water
Allende Meteorite	0.2[b]	0.004	0.36	0.008	0.097	0.20	0.034	20[c]
Murchison Meteorite	1.8	0.008	7.6	0.005	3.0	4.0	0.34	100[c]
DaG 476 (Mars)	-	0.017	0.92	0.019	1.0	0.38	0.064	-
EETA 79001 (Mars)	-	0.013	0.048	0.046	0.18	0.084	0.016	-
Lunar simulant (lava ash)	-	0.002	0.018	0.017	0.16	0.004	0.027	-
Agricultural soil	-	0.001	0.007	0.001	0.040	0.040	0.030	-
Total Contents								
Murchison Meteorite	18.6[c]	1.0[c]	32.4[c]	1.1[c]	13[c]	114[c]	0.28[c]	100[c]
Solar Abundance	3.94	0.95	0.40	0.008	0.068	0.76	0.0043	9.7[d]

a. Elements extracted under hydrothermal extractions in pure water at 120 °C for 15 minutes. Calcium (Ca), magnesium (Mg) and potassium (K) are extracted as elements in ionized forms Ca^{2+}, Mg^{2+} and K^+; sulfur (S) is extracted as sulfate SO_4^{2-}, nitrogen (N) as nitrate NO_3^- and phosphorus (P) as phosphate PO_4^{2-} (Mautner, 2002a).
b. Water-extractable carbon estimated as 10% of total organic carbon in Allende.
c. Total concentrations in Allende and Murchison (Jarosewich, 1971; Fuchs et al., 1973).
d. Calculated on the basis of the concentration of oxygen as 8.63 g/kg in solar material and $MW(H_2O/O) = 18/16$, considering oxygen as the limiting factor in constructing water. The solar abundance of hydrogen in solar matter is 789 g/kg which makes it not a limiting element for producing water.

Table A.3 Elements in biomass (gram element per kilogram dry biomass).[a]

	C	H	O	N	S	P	Ca	Mg	K
Bacteria	538	74	230	96	5.3	30	5.1	7.0	115
Brown algae	345	41	470	15	12	2.8	11.5	5.2	52
Plants (angio-sperm)	454	55	410	30	3.4	2.3	18	3.2	14
Mammals	484	66	186	128	16	43	85	1.0	7.5
Mammalian brain	484	66	186	99	6.7	12.2	0.32	0.55	11.6
Average biomass	462	59	329	67	7.4	15.5	21	3.0	34.4

a. From Bowen, (1966). The concentrations of C, H and O in brain tissue were not listed by Bowen and we assume them to be equal to concentrations in other mammalian tissue. Average biomass in the last row is the average elemental concentration in bacteria, brown algae, angiosperms, gymnosperms, mammalian tissue and mammalian brain biomass as given by Bowen.

Table A.4 Full wet biomass that can be constructed from each element in carbonaceous meteorites/asteroids and from a resource that contains elements with the cosmic elemental distribution[a,b] (gram biomass/kilogram resource).

	C	H	O	N	S	P	Ca	Mg	K
Carbonaceous Chondrites, Extractable Elements[c]									
Bacteria	13	108	123	0.34	5736	0.7	2353	2286	12
Mammals	15	110	125	0.25	1900	0.5	141	16000	81
Human brain	14	112	122	0.32	4537	1.7	37500	29091	117
Av. Biomass	16	112	119	0.48	4099	1.3	570	5260	40
Carbonaceous Chondrites, Total Elements[c]									
Bacteria	138	10	123	42	24453	147	10196	65143	108
Mammals	154	149	125	31	8100	102	612	456000	110
Human brain	149	97	122	40	19343	367	162500	829091	112
Av. Biomass	161	33	119	60	17474	284	2468	149918	112
Solar Distribution of Elements[c]									
Bacteria	29	7749	12	40	302	1	54	434	0.1
Mammals	33	7904	12	30	100	1	3	3039	2.3
Human brain	32	8024	12	38	239	3	861	5525	1.5
Av. Biomass	34	8055	12	57	216	2	13	999	0.5

a. Based on the data in Tables 1 and 2, and equation (A1) below. The wet biomass was calculated assuming a ratio of wet/dry biomass = 4.0, i.e., assuming that the content of each element except H and O in wet biomass is $c_{x,wet\ biomass} = 0.25c_{x,dry\ biomass}$. For H, the calculations used $c_{H,wet\ biomass} = [c_{H,dry\ biomass}/4 + (1,000 \times 2/18)]$ g/kg and for O, $c_{O,\ wet\ biomass} = [C_{O,\ dry\ biomass}/4 + (1000 \times 16/18)]$ g/kg.

b. To calculate the total biomass (in units of kilograms) that can be constructed from the extractable or total materials in 10^{22} kg asteroids, multiply the numbers in the top and middle Table, respectively, by 0.001×10^{22} kg. For the biomass constructed from 10^{26} kg cometary materials, multiply these data by 0.001×10^{26} kg.

TABLE A.5 Microbial and algal populations (in units of thousands) found per milliliter of solution on wetted meteorites that were inoculated by microorganisms from a peat bog.

	Clavi bacter Michiga nense	M. imper iale	Coryne bacteri um sp.	Chlo-Rella	Brown diatom algae	Blue-green fila-ment Algae	Total
Allende Meteorite	1,000	460	420	0	40	0	1,920
Murchison Meteorite	900	940	370	400	10	2	2,622
Sand reference	1,600	0	0	80	80	2	1,760

Appendix 2. Astroecology: Calculations

Appendix 2.1 Resources and Biomass

Living organisms take up nutrient elements from resource materials and incorporate them in biomass. Various types of resource materials contain various concentrations $c_{x,resource}$ of each nutrient element x, reported in Table A2.2 in units of g/kg. Similarly, $c_{x,biomass}$ (g/kg) is the concentration of element x in a given type of biomass as summarized in Table A2.3. Equation (A1) gives the amount of biomass, $m_{x,biomass}$ (kg) that could be constructed from an amount $m_{resource}$ (kg) of resource material if element x was the limiting factor and the other components of elements were available without limitation.

$$m_{x,biomass} = m_{resource}\, c_{x,resource}\, /\, c_{x,biomass} \qquad (A1)$$

Table A2.4 lists the amounts of biomass (kg) that can be constructed from each nutrient in 1 kg of the resource materials. Note that more than 1 kg of biomass could be constructed from 1 kg of resource materials as based on element x, for example, if a rare nutrient x is over-abundant in the resource materials.

Appendix 2.2 Rates of formation, steady-state amounts, and total time-integrated biomass.

We consider ecosystems where biomass $M_{biomass}$ (kg) is formed at a constant rate $dM_{biomass}/dt = k_{formation}$ (kg y^{-1}) and is used up or wasted at a rate $dM_{biomass}/dt = -k_{waste} M_{biomass}$ (kg y^{-1}). In astroecology, the formation may represent conversion of space resources to biomass while usage or waste may occur through leakage to space, or by the conversion of a fraction of the biomass to energy to provide power for the remaining biomass.

Note that the formation rate is zero order and the rate of waste is first order in $M_{biomass}$. Equation (A2) gives the rate of change of the biomass.

$$dM_{biomass}/dt = k_{formation} - k_{waste}M_{biomass} \qquad (A2)$$

Note that k_{waste} in units of y^{-1} represents the fraction of $M_{biomass}$ that is wasted per year. At steady state the rate of change of the biomass is zero e.g., $dM_{biomass}/dt = 0$, and equation (A3) gives the steady-state biomass $M_{biomass, equilibrium}$ (kg).

$$M_{biomass,\,steady\text{-}state}\,(kg) = k_{formation}\,(kg\,y^{-1})\,/\,k_{waste}\,(y^{-1}) \qquad (A3)$$

Next we calculate the time-integrated $BIOTA_{int}$ (Biomass Integrated Over Times Available) that exists in the ecosystem during a finite time period. In equation (A4) we integrate $M_{biomass,\,t}$ i.e., the biomass at any time, from the starting time t_o of the ecosystem to the end time t_f of life in the ecosystem.

$$BIOTA_{int} = \int_{t_o}^{t_f} M_{biomass,\,t}\,dt \qquad (A4)$$

After the formation of a given amount of biomass $M_{biomass,\,o}$ has been completed, it may be used up or wasted at the rate of $-k_{waste}$, i.e., the remaining amount of this unit of biomass decreases according to equation (A2) with $k_{formation} = 0$. The solution in equation (A5) gives the instantaneous amount that remains of this unit of biomass after time t.

$$M_{biomass,\,t} = M_{biomass,\,o}\,exp(-k_{waste}\,t) \qquad (A5)$$

By integrating equation (A5), we obtain the total integrated amount of this amount of biomass that will have existed from its formation to infinity.

$$BIOTA_{int} = M_{biomass,\,o}/k_{waste} \qquad (A6)$$

Note that equation (A6) applies to each unit of biomass that decays at the rate of $-k_{waste}M_{biomass}$ regardless of when it was formed. Therefore, the total integrated biomass of the ecosystem depends only on the total amount of biomass created and on the decay rate, but not on the rate of formation. Equation (A7) gives the total time-integrated biomass $BIOTA_{int,ecosystem}$ of the entire ecosystem. If the total amount of biomass created during the lifetime of the ecosystem is $M_{ecosystem}$ then

$$BIOTA_{int,\,ecosystem}\,(kg\text{-}y) = M_{ecosystem}\,(kg)\,/\,k_{waste}\,(y^{-1}) \qquad (A7)$$

Note that if waste is reduced to zero and no mass is lost from the biosystem then $k_{waste} = 0$ and the integrated $BIOTA_{int}$ is infinite for any finite amount of biomass. At the extreme, a single bacterium living forever would give an infinite amount of integrated biomass.

184

If all the mass $M_{resource}$ of the resource materials is converted to the maximum biomass that is allowed by the limiting nutrient according to Equation (A1), then equation (A8) gives the total time-integrated integrated $BIOTA_{int, ecosystem}$ of the ecosystem.

$$BIOTA_{int, ecosystem} = (M_{x,resource} \, c_{x,resource}/c_{x,biomass})/k_{waste} \qquad (A8)$$

An interesting case occurs if a fraction of the biomass is used to provide energy for the remaining biomass. Assume that the power requirement is $P_{biomass}$ (J s^{-1} kg^{-1}) and the energy yield is $E_{yield, biomass}$ (J kg^{-1}) per unit (kg) biomass converted to energy. If the biomass is converted to energy at the rate required to provide the needed power for the remaining biomass, then

$$(-dM_{biomass}/dt) \text{ (kg s}^{-1}) \; E_{yield, biomass} \text{ (J kg}^{-1}) =$$
$$P_{biomass} \text{ (J s}^{-1} \text{ kg}^{-1}) \; M_{biomass} \text{ (kg)} \qquad (A9)$$

This is similar to equation (A2) with a formation rate of zero and with $k_{waste} = P_{biomass}/E_{yield, biomass}$.
The remaining biomass after time t is given according to equation (A5) as

$$M_{biomass, t} = M_{biomass, o} \exp (-(P_{biomass}/E_{yield, biomass}) \, t) \qquad (A10)$$

The maximum energy can be obtained from a unit of mass by conversion to energy according to the relativistic relation $E = mc^2$. In this case $E_{yield, biomass} = c^2$ and assuming a power need of $P_{biomass} = 100$ Watt/kg, the decay rate of the biomass is

$$k_{use} = 100 \text{ (J s}^{-1} \text{ kg}^{-1}) / (3 \times 10^8)^2 \text{ (m}^2 \text{ s}^{-2}) =$$
$$1.11 \times 10^{-15} \text{ s}^{-1} = 3.5 \times 10^{-8} \text{ y}^{-1} \qquad (A11)$$

For a simple estimate of the amount of baryonic matter in the universe, the 10^{30} kg mass of the Sun may be multiplied by the 10^{11} stars and 10^{11} galaxies, yielding 10^{52} kg of baryonic matter. A more sophisticated calculation that was based on the volume of the universe (event horizon with a radius of 1.5×10^{26} m and volume of 1.4×10^{79} m^3) and the density of baryonic matter, 4.1×10^{-28} kg m^{-3} lead to a similar result of 5.9×10^{51} kg (Wiltshire, 2002).

If all the baryonic matter in the universe were converted to elements according to their proportions in biomass, then this process would yield 10^{52} kg of biomass. If a fraction of this biomass was converted to energy at the rate shown in equation (A11), then the initial rate of mass loss would be 3.5×10^{44} kg y^{-1}. There would be enough biomass left for one 50 kg human being after 3.3×10^9 years, and for a single bacterium of 10^{-15} kg after 4.4×10^9 years. The total time-integrated life will have been 2.8×10^{59} kg-y.

It is unlikely of course even in principle that all the matter in the universe can be brought together in one biosphere, since the galaxies are receding beyond their mutual event horizons. By analogous considerations, the duration of life using the 10^{41} kg matter in each galaxy is 2.6×10^9 years to the last human and 3.7×10^9 years to the last microbe. The total integrated life is 2.8×10^{48} kg-y per galaxy, which yields 2.8×10^{59} kg-y in all the galaxies. This amounts to 5.6×10^{57} human-years, or 5.6×10^{55} humans who will have each lived for 100 years. Although these numbers are not realistic and have large uncertainties, they illustrate the upper limits of biological and human life in the universe.

Appendix 2.3 Energy Flux, Temperatures and Habitable Zones about Stars

The luminosity of a star is equal to its power output, i.e., its energy output per unit time. The power output is related to the surface temperature according to

$$L \ (J \ s^{-1}) = 4\pi r_s^2 \sigma T^4 \tag{A12}$$

Here L is the luminosity, $4\pi r_s^2$ is the surface area and T ($^\circ$K) is the surface temperature of the star, and σ is the Stefan-Boltzmann constant, 5.67×10^{-8} J m^2 s^{-1} K^{-4}. The radius of the Sun is 6.96×10^8 m and its luminosity is 3.9×10^{26} J s^{-1}.

A spherical object with a radius r, at a distance R from the Sun, absorbs the solar flux intercepted by its projected area, at the rate

$$w_{abs} = (L/4\pi R^2) \ (\pi r^2) \ (1-a) \tag{A13}$$

Here the terms in the first parentheses give the solar energy flux at the distance R, the second parentheses the projected area, and the third parentheses account for the albedo, that is, reflection of part of the radiation.

The object also emits radiation depending on its radius and temperature according to equation (A12). At steady-state the absorbed and emitted radiation are equal and equations (A12) and (A13) can be combined to give the steady-state temperature as equation (A14).

$$T^4 = L(1-a)/16\pi R^2 \sigma \qquad\qquad (A14)$$

The steady-state temperature defines the habitable zone. It may be considered for example as the zone in which the steady-state temperature allows the existence of liquid water at 1 atmosphere total pressure, that is, 273 to 373 K (i.e., 0 to 100 C). More generally, we could define a narrower "comfort zone" say at 298 K (i.e., 25 C) or a wider "survivable zone" where life can exist with reasonable technology, for example, where the temperature ranges between 200 to 500 K.

We can use equation (A14) to calculate the habitable zone about the star as its luminosity changes. For example, it is estimated that in the red giant phase the luminosity of the Sun will increase to 2,000 times its present value. Then the habitable zone, for objects with an albedo of 0.3, will be between 21 and 39 au, and the comfort zone at 298 K at 33 au. At a later stage, the luminosity of the Sun may increase to 6000 times its current value. The habitable zone will then be from 36 to 68 au and the comfort zone will be centered about 57 au. Since the Sun will lose some mass and the planets will move out to larger orbits, Neptune, Pluto and the inner Kuiper Belt objects may move into or near the habitable zone at that time.

Pluto and the other Kuiper Belt objects are cometary nuclei that contain various ices. We may assume that objects that remain below 150 K will not evaporate, except for losing some highly volatile substances. Even at the hottest stage of the Sun, objects further than 225 au will remain at these low temperatures. This applies to the outer Kuiper belt which may extend to 1,000 au, and of course to the Oort cloud comets at 40,000 au, which will be only at 11 K even when the Sun is the most luminous. The preserved matter in the outer Kuiper Belt and the Oort Cloud may contain on the order of 10^{24} - 10^{25} kg of materials including organics and water that may be used as biological resources.

Note that travel to the 100 au comfort zone when the Sun is most luminous, at a reasonable speed of 10^{-4} c lasts only 16 years, and even to 1000 au to collect Kuiper Belt resources only 160 years. These resources can be accessed by humans with contemporary or slightly extended life-spans. In comparison, travel to the Oort clout at 40,000 au will last over 6,000 years and interstellar travel to nearby habitable stars will last from tens of thousands to millions of years. Such travel will require major changes in technology and biology. It is comforting therefore that humans like ourselves may survive in this Solar System through its hottest days, and that life can continue in our Solar System for inconceivably long hundreds of trillions of years during which the white dwarf Sun can sustain life.

References

1. Adams, F. and Laughlin, G., 1999. The Five Ages of the Universe, Touchstone Books, New York.
2. Anderson, J. M., 1981. Ecology for the Environmental Sciences: Biosphere, Ecosystems and Man, Edward Arnold, London.
3. Arrhenius, S., 1908. Worlds in Making, Harpers, London.
4. Baldauf, S. L., Palmer, J. D., and Doolittle, W. F., 1996. The Root of the Universal Tree and the Origin of Eukaryotes Based on Elongation Factor Phylogeny, Proc. Nat. Academy of Sciences, 93, pp. 7749-7754.
5. Began, M., Mortimer, M. and Thompson, D. J., 1996. Population Ecology: a Unified Study of Animals and Plants, Blackwell Science, Oxford.
6. Bowen, H. J. M., 1966. Trace Elements in Biochemistry, Academic Press, New York.
7. Brearley, A. J., and Jones, R. H., 1997. Carbonaceous Meteorites, in Planetary Materials, J. J. Papike, ed., Reviews in Mineralogy, Mineralogical Society of America, Washington, D. C. Vol. 36, pp. 3-1 - 3-398.
8. Crick, F. H. and Orgel, L. E., 1973. Directed Panspermia, Icarus 19, pp. 341-348.
9. Des Jardins, J. R., 1997. Environmental Ethics: An Introduction to Environmental Philosophy, Wadsworth, Belmont.
10. Davis, P. C. W., 1992. The Mind of God: The Scientific Basis for a Rational World, Touchstone Books, New York, pp. 194-198.
11. Dyson, F., 1979a. Time Without End: Physics and Biology in an Open Universe, Rev. Modern Phys. 51, pp.
12. 447-468.
13. Dyson, F., 1979b. Disturbing the Universe, Harper and Row, New York.
14. Dyson, F., 1988. Infinite in All Directions, Harper and Row, New York.
15. Dyson, F., 2001. Personal communication.
16. Fogg, M. J., 1993. Terraforming: A Review for Environmentalists, The Environmentalist 13, pp. 7-12.

17. Forward, R., 1988. Future Magic, Avon.
18. Fuchs, L. H., Olsen, E., and Jensen, K. J., 1973. Mineralogy, Mineral Chemistry and Composition of the Murchison (CM2) Meteorite. Smithsonian Contributions to the Earth Sciences 10, 1.
19. Gribbin, J., and Rees, M., 1989. Cosmic Coincidences., Bantam Books, New York, pp. 269-278.
20. Jarosewich, E., 1971. Chemical Analysis of the Murchison Meteorite, Meteoritics, 1, pp. 49-51.
21. Hargove, E. C. (Ed.), 1986. Beyond Spaceship Earth: Environmental Ethics and the Solar System, Sierra Club Books, San Francisco.
22. Gibbons, A., 1998. Which of Our Genes Make Us Humans, Science, 281, pp 1432-1434.
23. Hart, M. H., 1985. Interstellar Migration, the Biological Revolution, and the Future of the Galaxy, in Interstellar Migration and Human Experience, Ben R. Finney and Eric M. Jones, editors, University of California Press, Berkeley, pp. 278 – 291.
24. Hartmann, W. K., 1985. The Resource Base in Our Solar System, in Interstellar Migration and Human Experience, Ben R. Finney and Eric M. Jones, editors, University of California Press, Berkley, pp. 26-41.
25. Haynes, R. H., and McKay, C. P., 1993. The Implantation of Life on Mars: Feasibility and Motivation, Adv. Space. Res. 12, 133-140.
26. Heffernan, J. D., 1982. The Land Ethic: A Critical Appraisal, Environmental Ethics, 4, pp. 235-247.
27. Henderson, L. J., 1970. The Fitness of the Environment, Peter Smith, Glouster, pp. 312-318.
28. Hoyle, F., 1983. The Intelligent Universe, Michael Joseph, London, pp.218-224.
29. Lewis, J. S., 1996. Mining the Sky, Helix Books, Reading, Massachusetts.
30. Lewis, J. S., 1997. Physics and Chemistry of the Solar System., Academic Press, New York.
31. McKay, C. P., 1992. Does Mars Have Rights? An Approach to the Environmental Ethics of Planetary Engineering", in D. MacNiven, (Ed.), Moral Expertise, Routledge, New York, pp. 184-197.
32. McKay, C. P., Toon, O. B., Kasting, J. F., 1991. Making Mars Habitable, Nature, 352, pp. 489-496.
33. McKay, D. S., Carter, J. L., Boles, W. W., Allen, C. C., and Alton, J. H., 1993. JSC-1: A New Lunar Regolith Simulant, Lunar. Planet. Sci., 24, 963-964.
34. Mauldin, J. H., 1992. Prospects for Interstellar Travel, AAS Publications, Univelt, San Diego 1992.

35. Mautner, M., and Matloff, G. L., 1977. A Technical and Ethical Evaluation of Seeding the Universe. Bulletin American Astronomical Soc. 9, pp. 501.

36. Mautner, M. and Matloff, G. L., 1979. Directed Panspermia: A Technical and Ethical Evaluation of Seeding Nearby Solar Systems, J. British Interplanetary Soc., 32, pp. 419-423.

37. Mautner, M. N., 1995. Directed Panspermia. 2. Technological Advances Toward Seeding Other Solar Systems, J. British Interplanetary Soc. 48, pp. 435-440.

38. Mautner, M. N., 1997. Directed Panspermia. 3. Strategies and Motivation for Seeding Star-Forming Clouds, J. British Interplanetary Soc. 50, pp. 93-102.

39. Mautner, M. N., 1999. Formation, Chemistry and Fertility of Extraterrestrial Soils: Cohesion, Water Adsorption and Surface Area of Carbonaceous Chondrites: Prebiotic and Space Resource Applications, Icarus 137, pp. 178-195.

40. Mautner, M. N., 2002a. Planetary Bioresources and Astroecology. 1. Planetary Microcosm Bioassays of Martian and Meteorite Materials: Soluble Electrolytes, Nutrients, and Algal and Plant Responses, Icarus, 158, pp. 72-86.

41. Mautner, M. N., 2002b. Planetary Resources and Astroecology. Planetary Microcosm Models of Asteroid and Meteorite Interiors: Electrolyte Solutions and Microbial Growth. Implications for Space Populations and Panspermia, Astrobiology 2, pp. 59-76.

42. Morrison, D., 2001. The NASA Astrobiology Program, Astrobiology 1, pp. 3-13.

43. O'Keefe, S., 2002. Pioneering the Future, Syracuse University Lecture, April 12, 2002:. "To improve life here, to extend life to there, to find life beyond".

44. O'Leary, B. T., 1977. Mining the Apollo and Amor Asteroids, Science 197, pp. 363.

45. O'Neill, G. K., 1974. The Colonization of Space, Physics Today, 27, pp. 32-38.

46. O'Neill, G. K., 1977. The High Frontier, William Morrow.

47. Purves, W. K., Sadava, D., Orians, G. H., and Heller, H. C., 2001. Biology: The Science of Life, Sunderland, Massachusetts.

48. Ribicky, K. R.; Denis, C., 2001. On the final destiny of the Earth and the Solar System. Icarus, 151, pp. 130 - 137.

49. Rynin, N. A., 1971. K. E. Tsiolkovskii: Life, Writings, and Rockets, (Vol. 3, No. 7 of Interplanetary Flight and Communication. Leningrad Academy of Sciences of the U.S.S.R. Translated by the Israel Program

for Scientific Translations, Jerusalem, 1971). "The Earth is the cradle of the human mind, but one cannot live in the cradle forever".
50. Taylor, P. W., 1986. Respect for Nature: A Theory of Environmental Ethics, Princeton University Press, Princeton, pp. 45-68.
51. Wiltshire, D., 2002. Personal communication.

Appendix 3. Technical Papers (Full Reprints or Abstracts) On Astroecology, Space Resources and Directed Panspermia
Appendix A3.1

Astrobiology 2002, 2, 59-76

Planetary Resources and Astroecology. Planetary Microcosm Models of Asteroid and Meteorite Interiors: Electrolyte Solutions and Microbial Growth - Implications for Space Populations and Panspermia

Michael N. Mautner, Soil, Plant and Ecological Sciences Division, Lincoln University, Lincoln, New Zealand and Department of Chemistry, University of Canterbury, Christchurch 8002, New Zealand (mautnerm@lincoln.ac.nz)

ABSTRACT

Planetary microcosms were constructed using extracts from meteorites that simulate solutions in the pores of carbonaceous chondrites. The microcosms were found to support the growth of complex algal and microbial populations. Such astroecology experiments demonstrate how a diverse ecosystem could exist in fluids within asteroids and in meteorites that land on aqueous planets. The microcosm solutions were obtained by extracting nutrient electrolytes under natural conditions from powders of the Allende (CV) and Murchison (CM2) meteorites at low (0.02 g/ml) and high (10.0 g/ml) solid/solution ratios. The latter solutions, which simulate natural extractions of asteroids and meteorites by water during aqueous alteration, were found to contain >3 mol/L electrolytes and 1 mol/L organics, concentrated solutions favorable for pre biotic synthesis. The solutions and wet solids, inoculated with diverse microbial populations from a wetland, were found to support complex self-sustaining microbial communities for long periods (>8 months), with steady-state populations on the order of 4 x 10⁵ CFU/ml algae and 6 x 10⁶ CFU/ml bacteria and fungi. Planetary microcosm experiments based on meteorite materials can assist in assaying the fertilities of planetary materials and identifying space bioresources, targeting astrobiology exploration, modeling past and future space-based ecosystems, and evaluating sustainable populations in the Solar System. The results also suggest that protoplanetary nebulae can be effective nurseries for microorganisms and useful targets for directed panspermia.

Introduction

Carbonaceous objects in the Solar System include meteorites, asteroids, comets and interplanetary dust particles (IDPs). Under aqueous conditions, these objects form internal solutions that may originate and sustain microbial life. To assess these roles, it is necessary to understand the chemistry and biology of these materials. The present series of studies applies microcosm simulations, based on actual extraterrestrial materials in meteorites, to elucidate these properties (Mautner et al., 1995; Mautner 1997a; Mautner et al., 1997; Mautner, 2002; Mautner and Sinaj, 2002). The questions of interest are: What are the chemical properties of solutions formed when these materials are subjected to various aqueous environments? Can these solutions sustain complex microbial populations? The present study addresses these questions in relation to the potential roles of carbonaceous chondrites in early and future space ecosystems.

1. Organics on early planets

Meteorites, comets and in particular interstellar dust particles (IDPs) imported large amounts of organics to the early Earth, and presumably to Mars and the other planets (Delsemme 1995, Oro et al., 1995). The rate of infall of organic carbon to Earth during the intense bombardment period was of the order 10^8 - 10^9 kg yr^{-1} by IDPs, 10^5 - 10^6 kg yr^{-1} by comets and 10^3 - 10^4 kg yr^{-1} by meteorites (Chyba and Sagan, 1992).

2. Biogenesis on early planets

After infall to planets the IDPs, comets and meteorites can be exposed to water. The interiors of these objects are capable of forming concentrated solutions of organics and salts in the presence of mineral catalysts. Meteorite organics, as well as phosphate and other essential inorganic biological components such as Ca, Mg, Na, K, chloride and sulphate were shown to be extractable under planetary conditions (Mautner et al., 1995). The soluble meteorite organics include amino acids and adenine, as well as membrane-forming components and polycyclics that can affect energy conversion (Deamer, 1985; Deamer, 1992).

The interiors of IDPs (Kruger and Kissell, 1989; Maurette et al., 1995), cometary ponds (Clark, 1988), and the pores of meteorites on early

Earth (Mautner, 1997a; Mautner et al., 1997) could allow prebiotic synthesis and the origins of life. Of these objects, meteorite pores have the advantage of trapping the chemicals and allowing continuing chemical and microbial evolution. In this respect, it has been shown that various organics can be extracted under planetary conditions and some of these components can form vesicles (Mautner et al., 1995).

3. Biogenesis, lithopanspermia and directed panspermia in the Solar Nebula

Similar solutions can form in the interiors of carbonaceous chondrite parent asteroids during early aqueous alteration and in cometary nuclei during perihelion passes (Bunch and Chang, 1980; Tomeoka and Buseck 1985; Komle et al., 1991; Brearley and Jones 1998; Shearer et al., 1998). Both were suggested as potential sites for biogenenesis (Chyba and McDonald, 1995) and for transporting microorganisms (Hoyle and Wikramasinghe, 1978). Similarly, it was suggested that solar nebulae and young solar systems in star-forming interstellar clouds can be seeded with microorganisms, possibly using solar sailing or comets as vehicles (Mautner and Matloff, 1979; Mautner, 1995; Mautner, 1997b).

With respect to life in early solar systems, the survival of microorganisms on carbonaceous chondrite materials is of interest. In this respect, algae growing on meteorite dust in Greenland were observed as early as 1870 (Leslie, 1879; Maurette et al., 1986). The nutrient values of organic planetary materials were demonstrated on tholin, a synthetic analogue of organics formed under reducing Jupiter conditions (Stoker et al., 1990). Actual carbonaceous chondrite materials were examined in Murchison extracts were observed to support various soil microorganisms such as the oligotrophs *Flavobacterium oryzihabitans* and *Nocardia asteroides*, and experiments with *Pseudomonas fluorescens* showed that meteorite organics can serve as a sole carbon source (Mautner et al., 1995; Mautner et al, 1997). Indications were also found that the Murchison materials support the anaerobic thermophile eubacterium *Thermotoga maritima* and the aerobic thermophile *Thermus aquaticus* (H. W. Morgan, quoted in Mautner et al., 1997). In contrast, the Allende meteorite was observed to inhibit biological growth in some cultures (Mautner, 1997a). Recently, various carbonaceous chondrites were found to contain diverse microorganisms from terrestrial contamination (Steele et al., 2000). Plant tissue cultures of *Asparagus officinalis* and *Solanum tuberosum* (potato) can also uptake nutrients from the meteorite materials (Mautner, 1997a; Mautner et al., 1997).

4. Terraforming and space colonization.

In the future, asteroids may be used as resources for space colonization (O'Neill, 1974; O'Leary, 1977; Lewis, 1993). Carbonaceous chondrite materials from Phobos and Deimos may be used as fertilizers in Martian terraforming (Lewis, 1997). Relating to these applications, the Murchison CM2 meteorite was observed to have soil fertility parameters comparable to productive terrestrial soils (Mautner, 1997b; Mautner, 1999).

The microcosm simulations below will use realistic solid/solution ratios, and complex soil microbial communities that extend the previous work on isolated species (Mautner et al, 1995; Mautner et al, 1997).

Materials and Methods

1. Aqueous extractions

Solid samples of the Allende and Murchison meteorites were obtained from the Smithsonian Institute and from commercial sources. The mineralogies of the meteorites are well-established (Fuchs et al., 1973; Barber, 1981; Komacki and Wood, 1984). The samples were ground by hand in an agate mortar to achieve particle size distributions approximately similar to terrestrial soils, as reported elsewhere (Mautner and Sinaj, 2001). For example, the particle size distribution of Murchison powder is equivalent to silty clay soils with 57% clay-size particles ($<2\mu$), 41% silt-size particles, $2 - 20\mu$, and 2% sand-size particles $> 20\mu$. Samples of 80 - 200 mg of Allende powder and 40 - 80 mg of Murchison powder were placed in polythene tubes washed in 10% acetic acid for 24 hours to remove electrolyte impurities. Deionized water was added at various solid/solution ratios and the powders were extracted for four days at 20 °C with vortex shaking for one minute twice daily at the natural pH of 7.0 - 8.0 established by the powders. Extractions of several minerals showed that constant equilibrium concentrations are obtained in 2 - 8 day extractions under similar conditions (Mautner, 2001).

At solid/water ratios < 1, the suspended solids could be separated from the liquid by centrifuging and removing the liquid for analysis. However, at higher solid/water ratios the mixtures formed pastes that required special methods. The main method used for these samples was a rapid flush technique. These samples were extracted in 3 ml polythene syringes. After the extraction, twice 2 ml of deionized water was added to the syringes and pressed rapidly through a pre-washed filter. The flushing water was in contact with the extractant/solid paste for less than one minute, sufficient to dilute and remove the entrained extracts but not to dissolve significant amounts of further solutes. This assumption was tested by flushing non-extracted solid powders similarly. The small amounts dissolved from these non-extracted solids by the rapid flush were used as reference blanks.

As a control some of the samples were also analyzed by a different method, where a portion of the extracts in the pastes obtained at solid/water ratios of 1.0 and 2.0 was adsorbed on dry filter paper. The paper was weighed to determine the amount of extract adsorbed, which was approximately 50% of the total entrained extract. The paper was subsequently extracted into 4 ml of deionized water to analyze the solutes.

Trace metals were extracted from Allende and Murchison at $r_{solid/water}$ = 0.027 g/ml by 1M NH_4OAc solution, a standard soil extractant (Blakemore et al., 1987).

Anions in the extracts were analyzed by ion exchange chromatography using a Waters Ion exchange Chromatograph and Waters Baseline 810 software. The method used was Waters Ion Chromatography Method A-102 "Anion Exchange Analysis Using IC-Pak A HC Column Borate/Gluconate Eluent", with the samples filtered through a 0.1 micron filter paper prior to analysis. Cations were analyzed by a Shimadzu AA-6200 Atomic Absorption Flame Emission Spectrophotometer. Phosphate was analyzed by colorimetry using malachite green solutions (van Veldhoven and Mannerts, 1987). The uncertainty in the reported concentrations is estimated as ±30% from the results of replicate measurements. Similarly, the standard deviation of values obtained by the three different extraction methods for Allende at $r_{solid/water}$ = 1.0 was ±24%, and the average standard deviation of the constant c_{solid} values at $r_{solid/water}$ = 1 - 10 in Figures 3 and 4 below was ±18%. From these observations, the uncertainty in the data in Table 1 is estimated as ±30%.

2. Microbial cultures

A main objective of the present studies was to observe the development of mixed microbial populations in meteorite microcosms simulating asteroid and cometary interiors.

The main limitation of meteorite-based microcosm studies is the small amount of available materials. In this work, typically 20 mg was required per microcosm, and typical experiment series required several times this amount for various treatments and replicates. The minimum sizes of usable microcosms are defined by the requirement that the amounts of the chemicals and microorganisms should be sufficient to detect. Consideration of the current analytical methods and the usual range of extractable materials shows that extractions need to use 0.001 - 1 g mineral samples for anion and cation analysis, 0.01 - 0.1 g for phosphate analysis, 0.001 - 1 g for algal bioassays and 10^{-6} - 10^{-2} g for microbial studies (Mautner, 2001). Each of the present microbial cultures required 20 - 40 mg of materials.

Typically, our microcosm contains 0.2 - 1 ml extracts that allow repeated microbial population analysis. The small microcosms were prepared in 2 ml polythene microfuge tubes. In the present study two samples each of 20 mg of Allende and Murchison, and for reference acid-washed sand were extracted and sterilized in 1 ml of deionized water at 121 °C for 15 minutes, and inoculated with 20 microliters of the mixed microbial populations. These cultures will be denoted as "extracts". In another set of experiments, 100 mg of each solid was inoculated and wetted directly by 20 microliters of the inoculating solution. These cultures will be denoted as "wet solids". These extracts contained sufficient nutrients for microbial populations up to 10^8 CFU/ml on the "wet solids", in the post-log steady-state populations from 4 to > 31 days, if all the limiting nutrients would be utilized.

The inoculating solutions were chosen from a natural source with a mix of microorganisms that is expected to contain aerobes and anaerobes, autotrophs such as algae and heterotrophs adapted to humic or kerogen-like materials similar to those found in meteorites. For this purpose, samples of liquid and of wet soil from a local peat bog wetland reserve (Travis Swamp, Christchurch, New Zealand) was collected from the surface to a depth of one meter. The mixed sample from several layers was kept in a sealed jar terrarium allowing a slow diffusion of air for one year, creating a Winogradsky column (Winogradsky, 1949). During this time the soil differentiated into a 2 cm dark brown top layer, a 0.5 cm reddish brown middle layer and a 1.5 cm medium brown bottom layer. This layering may be due to microbial activity at various oxygen levels. The top level supported plant and algal growth indicating slowly exchanging aerobic conditions. The reddish brown layer may be due to sulphur or iron-oxidizing bacteria. In an oxygen restricted environment, the bottom layer is expected to be anaerobic or microooxic.

To produce inoculants for the aerobic cultures, 2 ml of wet soil was taken from the top layer and supplemented by 2 ml of algae cultures isolated from New Zealand soils and grown in algal nutrient cultures. The algal cultures included unicellular green algae *Chlorella sp.* and *Chlorosarcinopsis sp.*, filamentuos blue-greens *Leptolyngbya sp,.* and *Phhormidium sp.* and a gold-brown diatom *Navicula sp.,* which were identified microscopically. For anaerobic inoculants, 2 ml of wet soil was extracted from the bottom layer of the terrarium. The inoculating soil and algal material can be assumed to contain many bacteria and fungi, some of which were identified in the microcosm cultures as discussed below. The solids in the samples were washed three times with 8 ml of deionized water deareated with nitrogen, and centrifuged to separate the suspended solids and microorganisms. However, traces of oxygen could not be excluded and the experiments were suitable tested for facultative anaerobes.

No buffers were used, to avoid carrying over the buffer materials into the microcosms. Microorganisms used in the previous meteorite microcosms survived similar washing procedures without lysing (Mautner et al., 1997). Considering that <0.1 ml of water remained in the samples after each step, the procedure diluted all soluble components by a factor of 10^6 to remove soluble nutrients. Samples of 20 microliters of the final suspension of the washed solids, containing about 1 mg of the soil solids, were used for inoculations.

Following inoculation, the meteorite cultures were developed at 20 °C under natural light and dark cycles to allow the development of mixed populations including both photosynthetic autotrophs and heterotrophs. The culture vials were contained in sealed glass jars kept at 100% humidity to prevent drying. The jars were filled with air for the aerobic cultures and with a mix of 90% N_2 and 10% CO_2 for anaerobic cultures.

For monitoring the microbial populations, the cultures were plated on nutrient agar under aerobic or anaerobic conditions as appropriate, and on algal nutrient agar. Samples of colonies obtained from the mixed cultures were grown as isolates on separate plates and analyzed by Gram staining, oxidase and catalase responses and 96 well carbon source test plates (Biolog, Hayward, California).

Results

1. Aqueous extractions

As noted in the experimental section, the powders of Allende and Murchison were extracted in this work at solid/solution ratios < 1.0 by direct extraction and ratios > 2.0 by the flush or paper adsorption methods. For comparison, all three methods were applied to the Allende samples at solid/solution ratio of 1.0. For all the cations measured, the three methods gave values with standard deviation < ±30%, which shows that the consistency between the various methods is similar to the uncertainty of the replicate direct extractions themselves. The agreement amongst the methods is also supported by the absence of discontinuities in Figure 1 that include data points measured by all three methods.

The results of the extractions may be expressed in several related ways that reflect different physical quantities. The actual measurements yield aqueous concentrations in the extracts, c_{aq} (mg/l), which can be converted into the extractable content in the solid c_{solid} (mg/g) using equation (1).

$$c_{solid} \ (mg/g) \quad = \quad \frac{c_{aq} \ (mg/l) \ V_{aq} \ (ml)}{1,000 \ w_{solid} \ (g)}$$

$$= \quad \frac{c_{aq} \ (mg/l)}{1,000 \ r_{solid/solution} \ (g/ml)} \tag{1}$$

If the total extractable content at infinite dilution is also known the results can be converted to desorption isotherms, which will be presented elsewhere (Mautner et al., 2001). In relation to the microbial experiments, the relevant results are reported in terms of solution concentrations in Figure 1, that shows the values of c_{aq} as a function of the solid/solution ratio used in the extractions of the Allende meteorite. The extractions of Murchison yielded similar trends although at much higher absolute solute concentrations, that will be reported elsewhere (Mautner et al., 2001).

Figure 1 shows that the solute concentrations increase through the range of the solid/solution ratios applied. Table 1 reports the concentrations in the dilute solutions obtained at $r_{solid/solution}$ of 0.02 used for the microbial cultures. The Table also reports concentrations obtained at $r_{solid/solution}$ of 10.0 that simulate natural extractions by water in the meteorite and asteroid pores, at a porosity of 20% by volume. As $r_{solid/solution}$ increases by a factor of 500 between these extractions, the concentrations of most elements increases by factors of 200 - 500 over this range The results therefore suggest that most of the extractable electrolytes are present in the meteorite as soluble salts that dissolve fully even in the minimum amount of water used ($r_{solid/solution}$ = 10.0 g/ml). In the alternative case, if the concentrations had been controlled by adsorption/desorption equilibria on mineral surfaces, the extracted amounts c_{solid} would have increased significantly with increasing amounts of water. However, c_{solid} varied only moderately and became constant at $r_{solid/solution}$ = 1 - 10. These c_{solid} values can be calculated from the data in Table 1 using equation (1) and the reported c_{aq} values.

The extractable amounts of trace elements were also examined, using extraction by 1M NH$_4$OAc. The extractable amounts were (c$_{solid}$ (microgram/g), Allende, Murchison): B, 0.1, 0.5; Fe, 0.4, 9.0; Mn, 0.7, 11.0; Al, 0.06, 0.03; Cd, 0.0, 3.8; Cr, 0.0, 0.2; Cu, 0.004, 0.04; Ni, 85, 101, respectively.

Organic compounds constitute 18 mg/g of Murchison and 2.5 mg/g of Allende. Previous studies showed that 10% of the organic content is released under hydrothermal extraction at 121 °C, and about half as much may be extracted at 20° C (Mautner et al., 1995; Mautner, 1997a). The calculated amounts of organics released by aqueous extraction are about 1 mg/g organic C from Murchison. Assuming the same relations for Allende, 0.1 mg/g organic C is released by aqueous extraction.

2. Microbiology

The main culturable microorganisms that grew in the meteorite and sand extracts from the inoculates, and the steady-state populations that were established, are listed in Table 2. Two sets of cultures, in extracts and on wet solids, were grown in parallel. The development of the populations in the extracts is shown in Figures 2 - 4. Table 2 summarizes the steady-state populations obtained in both sets of experiments after 31 days, and the populations on the wet solids after 8 months of incubation.

Tentative identification of the observed species, obtained using Biolog carbon source plates, is listed in Table 2. Although the identifications are tentative, the tests provide useful information on the utilization of potential carbon sources in the meteorites, as discussed below. The similarity of our microorganisms to the species contained in the Biolog database is variable, with possibly reliable identification of *Eureobacterium saperdae* and *Pseudomonas putida*. The other identifications are tentative. However, all the species gave distinct colonies on nutrient agar which allowed counting the various microbial populations.

Several algae were also included in the cultures. The initial populations of the unicellular, filamentous and diatom species described above were, in the extract cultures, 10,000, 200 and 80 CFU/ml, respectively. Plate counts of the populations were obtained 15 days after inoculation, during the post-log phase of the bacterial and fungal populations. The algal populations in the extracts and on the wet solids are listed in Table 3. The algal populations increased in all the cultures after inoculation but reached smaller populations than the bacteria, as is also the case in terrestrial soil populations.

Microbial populations that were expected to contain anaerobes were grown in preliminary experiments under microoxic conditions. The inoculating cultures were obtained from the bottom layer of the wetland Winogradsky column described above. Samples of the cultures were plated on nutrient agar 27 days after inoculation and the plates were developed under microoxic conditions. Judging by colony morphology, the resulting populations resembled those in the aerobic samples. Under these conditions the main species remained *C. michiganense,* with smaller populations of the other original inoculating bacteria shown in Table 1. However, the late takeover by *Corynebacterium sp.* and the development of the yeasts and filamentuous fungi did not occur. The observations suggest that the bacterial species or strains isolated from the wetland are tolerant of microoxic conditions and can grow under such conditions on the meteorite extracts. We are investigating if true anaerobes can also grow on these materials.

Discussion

1. Applications of planetary microcosms

Microcosms are often used to simulate ecosystems under controlled conditions (Odum and Hoskins, 1957; Beyers, 1969). Meteorites allow studies, based on actual planetary materials, to simulate some aspects of planetary ecosystems that are not accessible otherwise.

The design of planetary microcosms depends on the objectives of the simulation, such as models of early life or future terraforming. Early life on Earth or in asteroids and comets would probably be anaerobic, while the aim of terraforming is to create habitable oxygen-rich environments. The latter aerobic systems are addressed primarily in this work.

The minimum usable sizes of microcosms are defined by the requirement of measurable amounts of chemicals and microorganisms. Current common analytical methods and the usual range of extractable contents require 0.001 - 1 g meteorite or mineral samples for nutrient analysis and for algal essays, and 10^{-6} - 10^{-2} g samples for microbial studies. A more detailed analysis is given elsewhere (Mautner, 2002). The present microbial cultures used 20 - 40 mg of materials per microcosm contained in solutions of 0.1 - 1 ml. Typical experimental series required several times this amount for various treatments and replicates.

2. Solution chemistry of asteroid and cometary interiors

Carbonaceous chondrite materials may be exposed to water in nature at widely varying solid/solution ratios. For example, meteorites that land in water are extracted at virtually infinite dilutions at very low solid/solution ratios. At the other extreme, water filling the pores of meteorites that fall on land are extracted by the penetrating water at a high excess of solid at solid/solution ratios of about 10 g/ml, given the porosity of about 20% by volume of CM2 meteorites (Corrigan et al., 1997).

The latter conditions also apply in asteroids during aqueous alteration when the internal pores are filled with water. A high concentration of soluble electrolytes in these fluids was implicit in our previous data (Mautner et al., 1995; Mautner, 1997a) and more recently in discussions by other authors (Bodnar and Zolensky, 2000; Cohen and Coker, 2000). The temperatures in these objects may range from 25 to 150 °C for thousands of years (Bunch and Chang, 1980; Tomeoka and Buseck, 1985; Brearley and Jones, 1998; Shearer et al., 1998).

Figure 1 and Table 1 illustrate that the equilibrium concentrations of ions in these solutions can vary widely with solid/solution ratio. Table 1 compares the concentrations of the extracted electrolytes from Allende and Murchison with terrestrial soil solutions. The results show that the concentrations in the dilute Allende extracts obtained a $r_{solid/solution}$ = 0.02 - 0.1 g/ml are generally below soil solutions or in the lower range of soil solutions. The concentrations of electrolytes in Murchison extracts at this solid/solution ratio are remarkably close to the median soil solution values except for higher levels of sulfate and phosphate.

For the solutions obtained at $r_{solid/solution}$ = 10 g/ml in Allende, the concentrations of cations and Cl exceed the upper limits of soil solutions by about an order of magnitude, while NO_3-N, SO_4-S and PO_4-P are in the range of terrestrial soil solutions. However, the concentrations of cations in the Murchison extracts at $r_{solid/solution}$ = 10 g/ml are higher than the upper range of soil solutions by over two orders of magnitude. Even the concentrations of NO_3-N and PO_4-P are comparable to the upper range of soil solutions. Sulfate is in large excess, with some possible implications discussed below. The high concentrations in these Murchison extracts may be in the toxic range for plants. The micronutrients and possibly toxic elements Mn and Fe are higher than the upper limit in typical surface soil solutions by factors of 55 and 360 respectively, and Ni is present at very high levels. However, some of these elements may also serve as oxidizable energy sources for microbial communities.

In molar units, the total concentration of the ions at $r_{sold/water}$ = 10.0 in Table 2 in the Allende extract is 0.097 mol/l with an ionic charge of

0.15eq/l, and in the Murchison extract 3.8 mol/l with an ionic charge of 6.6 eq/l. The latter values are much higher than the median electrolyte concentration of 0.031 mol/l and ionic charge of 0.034 meq/l in average surface soil solutions. The high electrolyte concentrations in the meteorite solutions imply that these solutions have high ionic strengths and osmotic pressures.

The amount of organic carbon that can be extracted from Murchison at 20 °C is about 1 mg/g (Mautner, 1997a). If this applies at $r_{solid/solution} = 10$ g/ml, the concentration of organic carbon in the asteroid fluids can be estimated as 10 g/l or about 1 mol/l in the pores of carbonaceous chondrites. This represents only 10% of the total organic carbon in Murchison, the rest remaining as insoluble compounds and organic polymer. If this insoluble carbon was present in the parent bodies originally as unpolymerized soluble material, the concentration in solutions in the parent asteroid or cometary fluids may have been in the range of 10 mol/l. The next section will discuss the prebiotic implications of these results.

3. Biogenesis in carbonaceous chondrite asteroids and meteorites

The preceding section suggests that the pores of meteorites and asteroids contain strong electrolytes with a total ionic concentration of >3mol/l, composed of the ions in Table 1; silicates not measured here; organic C of 1- 10 mol/l; pH 7 - 8 (Mautner, 1997a) and catalytic minerals including clay-like phyllosilicates with a large specific surface area of $3.7x10^4$ m^2/kg (Mautner, 1999). The dissolved metals and the clays present and minerals such as FeS can serve as catalysts (Bernal, 1951; Cairns-Smith and Hartman, 1986). The high concentrations and catalysts are conducive to complex organic synthesis, and the trapped products can undergo further reactions leading to large molecules.

The concentrations and osmotic pressures of these solutions can be comparable to that of a cell interior. Chemistry similar to that in cells can therefore occur in the meteorite/asteroid fluids without the need for enclosure in membranes. With such solutions an entire solution-filled meteorite or asteroid, or a solution layer in a comet, may function as a giant cell. This is also assisted by the relative concentrations of the N, P, K and soluble C, and in another group, Ca, Mg, Na, Cl and S, in these solutions that are similar to those in bacterial or algal biomass. Alternatively or later, primitive cells bound by inefficient or weak membranes may form (Deamer, 1985; Deamer, 1992) and survive without osmotic rupture under these conditions, dividing the solution into cells.

The asteroid environments also included CO_2 and H_2 captured from gases in the solar nebula, significant amounts of sulfur and sulfides (3% in Murchison) and temperatures of 25 - 150 °C in the mesophilic, thermophilic and hyperthermophylic range. These conditions are suitable for archaebacteria, possibly methanogens and sulphur bacteria. In other words, the conditions in the asteroids, or in similar solutions on meteorites that landed on early aqueous Earth and Mars are therefore consistent with the possible origins of life in the interiors of carbonaceous chondrite objects, or early adaptation to such environments. These conditions were at least as suitable as for the primitive microorganisms as hydrothermal vents. In addition, the asteroid and meteorite interiors had the advantage of trapping chemicals and microorganisms for thousands of years, allowing for continued chemical and biological evolution. Although meteorites were only a small fraction of the organics imported by dust particles, the 10^{14} kg meteorites landed in the first 10^8 years (Chyba and Sagan, 1992) could have accommodated 10^{10} kg biomass in 20^{24} microorganisms (see below), more than enough for a first evolving biota.

4. Indications of past microbial activity in asteroids

Several independent observations suggest that carbonaceous chondrites in our Solar System may have contained microbial life. For example, the Murchison organics contain amino acids and adenine and the polymer fraction resembles kerogen formed from biological materials. Microfossils in carbonaceous chondrites were reported by several observers (Claus and Nagy, 1961; Urey, 1962; Hoover, 2001). Life on Earth originated soon after the period of late heavy bombardment suggesting that microorganisms may have been delivered to Earth by meteorites, asteroids or comets.

Some of our observations are consistent with possible biological activity in the carbonaceous chondrite parent bodies. We found that the composition of the Murchison materials resembles biologically developed terrestrial soils in its overall organic content, C/N ratio, cation exchange capacity and concentrations of the available macronutrients (Mautner 1997b; Mautner, 1999). The effects of the Murchison meteorite on microorganisms are also similar to the effects of biologically developed soils (Mautner, 1997b; Mautner et al., 1997). As noted above, the ratios of soluble N, P, K and C, and those of soluble Ca, Mg, Na, Cl and S in the meteorite are also remarkably similar to those in biological materials. This similarity may suggest that the soluble elements were deposited from a microbial biomass. Alternatively, it may suggest that the present biological elemental ratios reflect the conditions of early life in carbonaceous chondrite asteroids or meteorites.

A further indication of possible microbial origin may be provided by the large concentration of soluble sulfate, 9.4 mg/g, observed in Murchison (Table 1 and Mautner, 1997a). This is much higher than in any other type of meteorite, including stony and Martian meteorites, igneous terrestrial analogues and serpentine (Mautner et al., 2001), where the soluble sulfate concentration is only 0.01 - 0.1 mg/g. However, the Murchison parent body was formed under reducing Solar Nebula conditions in the absence of oxygen, and it is uncertain if oxidised sulfur can be formed chemically under these conditions. It may have required sulfide-oxidising bacteria to oxidise sulfur under these conditions. We are testing this hypothesis by examining the SO_4-S isotopic composition in carbonaceous chondrites for biological signatures.

In total, the observed chemistry and microbiology are consistent with the possible origin and past biology in carbonaceous chondrite objects. Implications for future life in space will be discussed in the next sections.

5. Microbial populations in meteorite solutions

In extension of the previous studies of pure microbial populations (Mautner, 1997a; Mautner et al., 1997), the present experiments used complex microbial inoculants. The results in Tables 2 and 3 show that complex microbial communities including bacteria, fungi and algae can grow on the meteorites. Table 2 shows that comparable population densities were observed in the dilute extracts and in the liquid on the wet solids.

Figures 2 and 3 and Table 2 show that the extracts of the two meteorites developed comparable microbial populations, with the Allende populations being slightly lower possibly because the lower concentration of nutrients in these extracts. On the other hand, Allende yielded the most diverse microbial populations, and in the long term its populations reached comparable levels to those in the Murchison solutions. Several additional microorganism species, not shown in Table 2, were also observed in Allende in small numbers. However, microorganisms in the extracts of both meteorites exhibited overall similar growth profiles, including the replacement of *Clavibacter michiganense* by *Corynebacterium sp.* as the dominant species after about 15 days.

The long-term survival of microorganisms on these materials was tested by measuring the populations after 8 months of incubation. Table 2 shows that the populations survived and in fact increased during this long period. The main species *C. michiganense, M. imperiale* and *Corynebacterium sp.* reached practically identical populations in both the Allende and Murchison cultures. Both also showed large populations of an additional unidentified species that form yellow globular colonies, and some additional new species in smaller numbers. The total population counted after 8 months in both cultures was also practically identical, 4.7 - 4.8 x10^6 CFU/ml.

The general similarity of microbiology in the two meteorites is notable considering the much higher concentration of electrolytes in the Murchison extracts. Given that the microorganisms grew and survived also on nutrient-free acid-washed sand, these species may be oligotrophs possibly living on organics from the laboratory air. We demonstrated in the past that Murchison organics can provide the sole carbon source for microorganism (Mautner et al., 1997). The long-term survival of microorganisms on the wet meteorite solids after eight months of incubation shows at least that the meteorite solutions do not contain toxic components that would prevent microbial growth.

The total algal populations in the Allende extracts were also comparable to that in the Murchison extracts as seen in Table 3, while in the concentrated solutions on the wet solids Allende gave lower algal populations than Murchison, and even than wet sand. This suggests that the concentrated Allende solutions inhibited algal growth. Similar inhibitory effects of Allende extracts were observed on potato tissue cultures (Mautner, 1997a).

Both meteorites supported more diverse microbial populations than inert sand. The meteorite extracts also showed the development of *Corynebacterium sp.* as the dominant species after about 15 days. The large populations of *C. michiganense* and *P. putida* in the extracts of inert sand suggests that these microorganisms are oligotrophs. The fungi in later stages of the cultures may utilize the biomass produced by autotrophs and the detritus from the bacterial and algal populations.

The biomass that can be supported by any given nutrient x in the solid soil or meteorite can be calculated by equation (2).

$m_{biomass,xl}$ (g) =

$$1,000 \ c(x)_{solid} \ (mg/g) \ m_{solid} \ (g) \ / \ c(x)_{bioamass} \ (\mu g/g) \quad\quad (2)$$

Here $m_{bioamss,xl}$ is the amount of biomass that can be obtained if limited by nutrient x; $c(x)_{solid}$ is the concentration of extractable x in the solid, m_{solid} is the mass of the solid extracted, and $c(x)_{biomass}$ is the concentration of bioavalable x in the dry biomass.

For calculating the limiting microbial population in the aqueous extract of the solid, the amount of microbial biomass in the solution can be expressed by equation (3).

$$m_{biomass,aq} \text{ (g)} = c_{microorganisms} \text{ (CFU/ml) } V_{aq} \text{ (ml) } m_{microorganism} \text{ (g)} \quad (3)$$

Here $m_{biomass,aq}$ is the amount of biomass in V_{aq} volume of solution, contained in organisms each of $m_{microorganism}$ dry biomass. Combining equations (2) and (3) yields the limiting microbial population density in solution.

$$c_{microorgansims(aq),xl} \text{ (CFU/ml)} = $$

$$\frac{1,000 \; c(x)_{solid} \text{ (mg/g) } r_{s/w} \text{ (g/ml)}}{c(x)_{biomass} \text{ (µg/g) } m_{microorganism} \text{ (g)}} \quad (4)$$

Here $c_{microrganism(aq),xl}$ is the concentration of microorganisms allowed if x is the limiting nutrient, $r_{s/w}$ is the solid/solution ratio in the extraction and $c(x)_{solid}$ is the concentration of nutrient x in the solid that is extractable at this solid/solution ratio.

The last two rows of Table 1 show the sustainable populations of typical bacteria with a radius of 1 micron and dry mass of 2×10^{-12}g calculated using equation (4). The results show for example that the Ca content in the concentrated Allende solutions (at $r_{solid/solution} = 10$ g/ml) can sustain 4.1×10^{10} bacteria/ml, while the NO_3-N content in the extract is sufficient only for 3.5×10^8 bacteria/ml. These calculations show that NO_3-N, PO_4-P, K, and even soluble C limit the bacterial populations to quite comparable levels of $3.5\text{-}10 \times 10^8$ bacteria/ml in the concentrated Allende extracts. The nutrients K and P lead to similar limiting levels also in the Murchison extracts, consistent with the similarity of the populations observed in Table 2 in the two extracts. The limiting nutrients in the meteorite extracts are nitrate, phosphate and potassium, that are also typical limiting nutrients in many terrestrial ecosystems.

The other essential nutrients Ca, Mg, Na and SO_4-S are all sufficient to provide larger populations over 10^{10} bacteria/ml in the Allende and over 10^{12} bacteria/ml in the Murchison solutions.

Note that soluble elements in the group N, P, K and C and elements in the Ca, Mg, Na, Cl and S in the meteorite extracts can support mutually similar populations of bacterial biomass. This is a consequence of equation (4) as $c(x)_{solid}$ and $c(x)_{biomass}$ are similar in these groups of elements. In other words, the relative concentrations of the soluble forms of these elements in Murchison are similar to their relative amounts in bacterial biomass.

The high concentrations of sulfate and Ni, and possibly Fe and Mn, may be toxic to some organisms. However, we did not observe toxic effects, as the microbial populations in the concentrated solutions on the wet solids were comparable to or larger than in the more dilute solutions (Table 2).

The Biolog tests provide some useful information on the possible meteorite components utilized by the microorganisms. All of the microorganisms can use glycerol, a polyalcohol that may be present in Murchison considering the presence of various alcohols and other hydroxylated compounds such as carboxylic and amino acids. One of the microorganisms, *Pseudomonas putida,* also utilizes other Murchison components such as acetic acid and alanine. Murchison also contains a large number of other likely nutrients for heterotrophs that were not included in the Biolog tests.

6. Sustainable populations on asteroid resources

The measured amounts of bioavailable nutrients allows and experiment-based estimate of the biomass that could have existed in the early Solar System or that could be established in the future, based on carbonaceous chondrite resources. The biomass that could be obtained as limited by nutrient x is obtained from equation (2), using $m_{solid} = 10^{22}$ kg as the total mass of the carbonaceous asteroids (Lewis, 1997) and the allowed microbial population is obtained from $m_{biomass,xl}/m_{microorganism}$, where $m_{microrganism} = 2 \times 10^{-12}$ g is used as an estimated average. Further, the allowed human population can be calculated by assuming, for example, 10^4 kg biomass supporting a human on a terraformed planet or space colony.

$$n_{population} = 10^{-4} \, c(x)_{\;solid} \, (mg/g) \, m_{solid} \, (g) \, / \, c(x)_{biomass} \, (\mu g/g) \quad (5)$$

Given the similarity of equations (4) and (5), the allowed biomass (kg) and human population, based on carbonaceous chondrite Murchison CM2 type asteroids, is obtained by multiplying the last row in Table 2 by 2×10^9 and 2×10^5, respectively. This of course yields again NO_3-N and PO_4-P as the liming nutrients. Both lead to similar values, with an NO_3-N limited biomass of 1.3×10^{18} kg in 6.5×10^{32} microorganisms, supporting a human population of 1.3×10^{14}, or about 10^{-4} kg biomass per kg asteroid material. The values based on PO_4-P are similar, being larger by a factor of 1.4. Note that the actual populations in Table 2 are lower by about a factor of 100 than the calculated nutrient-limiting populations. The asteroid-based populations would be lower by a factor of 100 using these actual observed populations.

6. Natural and directed panspermia

Comets were proposed as vehicles for natural panspermia (Hoyle and Wikramashinge, 1978). As noted above, comets may contain pockets of water at 20 - 100 °C during perihelion passes, containing fluids similar to the concentrated Murchison extracts, that support microbial growth. Comets may be also used deliberately for directed panspermia, by seeding them with microorganisms that can grow to a substantial biomass during perihelion passes. The microorganism bearing comets can then be fragmented and propelled or allowed to eject naturally to interstellar space (Mautner, 1997b).

However, comets may have limited potential as liquid water may not form, or may be lost rapidly by evaporation. This also limits the surface temperatures to <180 K, too low for microorganisms (Komle et al., 1991; Lewis, 1997). Liquid water may exist at best for short periods, in small pockets shielded by carbonaceous deposits. These conditions are not favorable for complex chemistry, and do not allow sustained evolving microbial populations.

In contrast, asteroid interiors during aqueous alteration can contain large volumes of water for thousands of years. For example, a 10 km radius asteroid or cometary nucleus of about 10^{16} kg with 10% water content can contain over 10^{15} liters of nutrient electrolyte solutions similar in composition to those in Table 1. Following the preceding section, this asteroid may accommodate a biomass of 10^{12} kg in 5×10^{26} bacteria and about 10^{24} algae. As noted, the total asteroidal mass of 10^{22} kg can accommodate over 10^{32} bacteria and 10^{30} algae.

In fact, large numbers of asteroids undergo aqueous alteration simultaneously. Collision amongst these objects are frequent (Lewis, 1997), and impact collisions can distribute microorganisms (Mileikowsky et al., 2000). Even small meteorites ejected by these collisions can provide enough shielding to protect microorganisms in space for months (Horneck et al., 1994), until recapture by another object. These considerations, combined with the present results, suggest that significant microbial populations can grow in aqueous asteroids. Collisions can distribute the microorganisms efficiently in the early Solar System.

Part of this population will be preserved at low temperatures in the interiors of asteroids and comets, and a fraction subsequently delivered to planets. Other asteroids and comets are ejected to interstellar space and some of these can spread the microorganisms to other protoplanetary nebulae, where they can multiply similarly and propel to yet further nebulae. This mechanism therefore proposes asteroids, protoplanetary nebulae and early solar systems for the growth and dispersion of microbial life, instead or in addition to the cometary proposals (Hoyle and Wickramasinghe, 1978).

A similar mechanism can be applied in directed panspermia (Crick and Orgel, 1973; Mautner and Matloff, 1979; Mautner, 1995; Mautner, 1997b). For example, comets may be seeded with microorganisms. Natural melting during perihelion passes may allow the microorganism to multiply. Microbial inoculants may be also inserted deeper into the cometary nuclei together with artificial heat sources to melt the ices and create sub-surface pools. Given the possible relations between carbonaceous asteroids and comets, concentrated solutions in these pools may be similar to those observed in the Murchison extracts, and can allow similarly the growth of large microbial populations. Eventually, the comet may be fragmented and ejected into interstellar space toward new Solar Systems in star-forming clouds, carrying the microbial payload. If non-periodic comets in parabolic orbits are seeded, this ejection will occur naturally.

The cometary interiors can shield the microbial content from prolonged space radiation, although the effects over transit times of millions of years are not known. If long-lived radioactive heat sources are used, a self-recycling microbial community can renew itself genetically during the long interstellar flights. The microorganisms can multiply and disperse collisionally in the target protoplanetary nebulae and asteroids, seed local planets and eventually disperse by comets to further Solar Systems. Using the above relations between nutrients and biomass, the 10^{25} kg of comets in the Oort cloud can yield 10^{21} kg biomass comprised of over 10^{35} microorganisms, sufficient to seed new Solar Systems throughout the entire Milky Way galaxy (Mautner and Matloff, 1979; Mautner, 1995; Mautner, 1997b). It is also possible that intelligent civilizations evolving in the seeded habitats will propagate life further in the galaxy deliberately.

7. Terraforming and space agriculture

Carbonaceous chondrites are likely to be the main sources of carbon and water in space-based agriculture. In space colonies and in terraformed asteroids, they may be used as soils and fertilizers. The carbonaceous moons Phobos and Deimos may be mined for soils and fertilizers in Martian terraforming.

Algae are likely to be used as colonizing microorganisms (Friedmann and Ocampo-Friedmann, 1995). A viable soil ecology will subsequently require a diverse microbial population and the recycling of nutrients by bacteria and fungi. The microbial results above demonstrate that carbonceous chondrite soils can sustain diverse microbial ecosystems.

If ground into particles similarly to the present studies, the Murchison soil will be similar in particle size distribution to silty clay. In this form the agriculturally useable moisture content will be between the wilting point at 20% w/w and field capacity at 40% w/w. These moisture contents are in the range used in the extractions above and will yield nutrient ion concentrations similar to those of the concentrated solutions in Table 1. The dilute solutions in Table 1 may be also used for hydroponics. Using the figures in the preceding section, the total asteroid materials used as synthetic soils allow a biomass of 10^{18} kg and a human population of 10^{14} in these terraformed colonies.

Summary and Conclusions

The previously reported studies on carbonaceous chondrites (Mautner, 1997a; Mautner et al, 1997) have been extended here to natural aqueous conditions of high solid/solution ratios. The main experiment-based conclusions are:

1. Planetary microcosms, based on actual extraterrestrial materials in meteorites, are useful tools in experimental astroecology.
2. Based on microcosm studies, the interiors of carbonaceous chondrite meteorites, asteroids and comets can contain highly concentrated solutions of electrolytes, nutrients and organics. *(Observation: >3 mol/l electrolytes, 1 - 10 mol/l organics)*
3. In the presence of mineral catalysts, these trapped concentrated solutions are suitable for prolonged, stepwise synthesis of complex organics.
4. The interiors of asteroids during aqueous alteration, or meteorites landed on aqueous planets, are therefore suitable for potential biogenesis.

5. The resulting indigenous, or introduced, microorganisms can grow in the interior solutions in meteorites and asteroids.

6. Possible microfossils, the biomass-like ratios of macronutrients and high sulfate content may suggest past biological activity in carboanceous chondrite asteroids. *(Indicative observations: Soil fertility properties of Murchison are similar to biologically developed soils; organic polymer similar to coal; ratios of soluble N, P, K and C, and Ca, Mg, Na, Cl and S are comparable to those in bacterial biomass; high sulfate content in Murchison).*

7. Complex recycling communities of algae, bacteria and fungi develop and survive for substantial periods in these solutions. *(Observation: Algal populations of $>10^5$ and microbial populations $>10^6$ CFU/ml of six species surviving over 8 months on wet Allende and Murchison).*

8. Carbonaceous asteroids containing nutrient solutions can distribute micoorganisms during a period of collision-mediated panspermia in the Solar Nebula and in the early Solar System. *(Available nutrients allow a biomass of 10^{18} kg in a population over 10^{32} microorganisms in the asteroid belt.)*

9. Similarly, comets can be used as vehicles, and protoplanetary nebulae can be used as targets and incubators in directed panspermia missions for seeding new planetary systems with microbial life *(The nutrients in the Oort belt comets allow a biomass of 10^{21} kg containing 10^{35} microorganisms, sufficient to seed all new solar systems in the galaxy).*

10. Carbonaceous chondrites are suitable soil resources for planetary terraforming and space colonization. *(Based on the limiting nutrients NO_3-N and PO_4-P, the total asteroid material can support a population of 10^{14} humans).*

The present microcosms examined separately the nutrient contents and the microbial populations in the microcosms. In real ecosystems, microbial activity and available nutrients are interdependent, and should be monitored simultaneously.

We are extending our planetary microcosm studies to further types of meteorites representing other asteroids, also to Martian meteorites and simulants (Mautner, 2002), plausible planetary conditions such as CO_2 atmospheres, and anaerobic microbial populations. These microcosm simulations will model the complex interactions between nutrients, pH, temperature, light flux, and biological populations in planetary ecosystems. For example, the first ranking of carbonaceous and Martian meteorites, based on extractable nutrients, and algal and plant yields, resulted in a ranking of fertilities as Martian basalts > terrestrial basalt > carbonaceous chondrites, lava ash > cumulate igneous rock. (Mautner, 2002). Eventually, these microcosm studies can help in targeting astrobiology exploration; identifying bioresources in the Solar System; and modeling ecosystems for terraforming, space colonization and directed panspermia.

Acknowledgements

I thank Ms. Catherine Trought for assistance with the aqueous extractions and analysis, Dr. Paul Broady for the algal samples, Dr. Mark Braithwaite for microbial identification and Dr. Eric Forbes for a review of the manuscript, and the Smithsonian Institution for a gift of Allende and Murchison samples. This work was supported by grant 99-LIU-014 ESA from the Marsden Fund administered by the Royal Society of New Zealand.

REFERENCES

1. Barber, D. J. (1981) Matrix phyllosilicates and associated minerals in CM2 carbonaceous chondrites. *Geocim. Cosmochim. Acta* 45, 945-970.
2. Bernal, J. D. (1951) The Physical Basis of Life. Rouletge Kegan Paul, London.
3. Beyers, R. J. (1969) The pattern of photosynthesis and respiration in laboratory microecosystems. In *Primary Productivity in Aquatic Environments,* C. R. Goldman, ed. Mem. Ist. Ital. Idrobiol., 18 Suppl., U. California Press, Berkely, pp. 62-74.
4. Blakemore, L. C., Searle, P. L., and Daly, B. K. (1987). *Methods for Chemical Soil Analysis,* New Zealand Soil Bureau Scientific Report 80. Department of Scientific and Industrial Research, Lower Hutt, New Zealand.
5. Bodnar, R. and Zolensky, M. (2000). Liquid-water fluid inclusions in chondritic meteorites: Implications for near-surface P-T conditions on parent asteroids. *Journal of Conference Abstracts* 5, 223.
6. Bowen, H. J. M. (1966) *Trace Element in Biochemistry.* Academic Press, New York, pp. 31,32.
7. Brearley, A. J. and Jones, R. H. (1998). Carbonaceous meteorites. In *Planetary Materials* (J. J. Papike, ed.) Reviews in Mineralogy v. 36, Mineralogical Society of America.
8. Bunch, T. E. and S. Chang (1980). Carbonaceous chondrites. II. Carbonaceous chondrite phyllosillicates and light element

geochemistry as indicators of parent body processes and surface conditions. *Geochim. Cosmochim. Acta* 44, 1543-1577.

9. Cairns-Smith, and Hartman, H. (1986) *Clay Minerals and the Origins of Life.* Cambridge University Press, Cambridge, UK.

10. Chyba, C. F., and C. Sagan (1992) Endogenous production, exogenous delivery and impact-shock synthesis of organic molecules: An inventory for the origins of life. *Nature* 335, 125-132.

11. Chyba, C. F. and G. D. McDonald (1995) The origin of life in the Solar System: Current issues. *Annual Rev. Earth Planet. Sci.* 23, 215-249.

12. Clark, B. C. (1988) Primeval procreative comet pond. *Orig. Life Evol. Biosphere* 18, 209-238.

13. Claus, G. and Nagy, B. (1961) A microbial examination of some carbonaceous chondrites. *Nature* 192, 594-596.

14. Cohen, B. A. and Coker, R. F. (2000) Modelling of liquid water on CM meteorite parent bodies and implications for amino acid racemization. *Icarus* **145,** 369-381.

15. Corrigan, C. M.; Zolensky, M E.; Dahl, J.; Long, M.; Weir, J.; Sapp, C.; Burkett, P. J. (1997) The porosity and permeability of chondritic meteorites and interplanetary dust particles. *Meteoritics and Planetary Science* 32, 509 - 515.

16. Crick, F. H. and Orgel, L. E. (1973) Directed panspermia. *Icarus* 19, 341-346.

17. Deamer, D. W. (1985) Boundary structures formed by organic components of the Murchison carbonaceous chondrite. *Nature* 317, 792-794.

18. Deamer, D. W. (1992) Polycyclic aromatic hydrocarbons: Primitive pigment systems in the prebiotic environment. *Adv. Space Res.* 12, 183-189.

19. Delsemme, A. H. (1995) A cometary origin of the biosphere: A progress report. *Adv. Space Res.* 15, 49-57.

20. Friedmann, E.I. and Ocampo-Friedmann, R. (1995). A primitive cyanobacterium as a pioneer microorganism for terraforming Mars. *Adv. Space Res.* 15(3), 243-246.

21. Fuchs, L. H.; Olsen, E.; and Jensen, K. J. (1973) Mineralogy, mineral-chemistry, and composition of the Murchison (C2) meteorite. *Smithsonian Contributions to the Earth Sciences,* number 10. pp. 1-39. Smithsonian Institution Press, Washington, D. C.

22. Hoover, R. B. (2001) Personal communication.

23. Horneck, G.; Bucker, H.; Reitz, G. (1994) Long-term survival of bacterial spores in space. *Adv. Space Res.* 14, 41-45.

24. Hoyle, F.; Wikramasinghe, C. (1978) *Lifecloud: the Origin of Life in the Universe.* J. M. Dent and Sons, London.

25. Komacki, A. S., and Wood, J. A. (1984) The mineral chemistry and origin of inclusion matrix and meteorite matrix in the Allende CV3 chondrite. *Geochim. Cosmochim. Acta* 48, 1663-1676.

26. Komle, N. I.; Steiner, G.; Baguhl, M.; Kohl, H.; and Thiel, K. (1991) The effect of non-volatile porous layers on temperature and vapor pressure of underlying ice. *Geophys. Res. Lett.* 18, 265-268.

27. Kruger, F. R. and Kissel, J. K. (1989) Biogenesis by cometary grains - thermodynamic aspects of self-organization. *Orig. Life Evol. Biosphere* 19, 87-93.

28. Leslie, A. (1879) The Arctic Voyages of Adolf Eric Nordenskiold. McMillan and Co., London. P. 65. Refers to observations of algae by the botanist Dr. Berggren on cryoconite, later identified as micrometeorite dust (Maurette et al. 1986).

29. Lewis, J. S. (1993) *Resources of near-Earth space.* University of Arizona Press, Tucson, Arizona.

30. Lewis, J. S. (1997) *Physics and Chemistry of the Solar System.* Academic Press, New York.

31. Maurette, M., Hammer, C., Brownlee, D. E., Reeh, N. and Thomsen, H. H. (1986) Placers of cosmic dust in the blue ice lakes of Greenland. *Science* 233, 869-872.

32. Maurette, M., Brack, A., Kurat, G., Perreau, M. AND Engrand, C. (1995) Were micrometeorites a source of prebiotic molecules on the early Earth? *Adv. Space Res.* 15, 113-126.

33. Mautner, M. N. (1995) Directed panspermia. 2. Technical advances toward seeding other solar systems and the foundations of panbiotic ethics. *J. British Interplanet. Soc.* 48, 435-440.

34. Mautner, M. N. (1997a) Biological potentials of extraterrestrial materials. 1. Nutrients in carbonaceous meteorites, and effects on biological growth. *Planetary and Space Science* 45, 653-664.

35. Mautner, M. N. (1997 b) Directed panspermia. 3. Strategies and motivation for seeding star-forming clouds. *J. British Interplanetary Soc.* 50, 93-102 (Also available at www-panspermia-society.com).

36. Mautner, M. N. (1999) Formation, chemistry and fertility of extraterrestrial soils: Cohesion, water adsorption and surface area of carbonaceous chondrites. Prebiotic and space resource applications. *Icarus* 137, 178 – 195.

37. Mautner, M. N. (2002) Planetary resources and astroecology. 1. Planetary microcosm bioassays of Martian and meteorite materials: nutrients, and plant and algal responses. Submitted for publication.

38. Mautner, M. N. and Matloff, G. L. (1979) Directed panspermia: A technical and ethical evaluation of seeding nearby solar systems. *J. British Interplanetary Soc.* 32, 419-423.

39. Mautner, M. N.; Leonard, R. L.; and Deamer, D. W. (1995) Meteorite organics in planetary environments: Hydrothermal release, surface activity and microbial utilization. *Planet. Space Sci.* 43, 139-147.

40. Mautner, M. N., A. J. Conner, K. Killham and D. W. Deamer (1997) Biological potential of extraterrestrial materials. 2. Microbial and plant responses to nutrients in the Murchison carbonaceous meteorite. *Icarus* 1997, 245-253.

41. Mautner, M. N. and Sinaj, S. (2002) Water-extractable and exchangeable phosphate in Martian and carbonceous meteorites and in planetary soil analogues. Submitted for publication.

42. Mautner, M. N., Cameron, K. C.; and Trought, K. (2002) Planetary resources and astroecology. 3. Water-extractable ions and nutrients in carbonaceous and stony meteorites. In preparation.

43. Mileikowsky, C., Cucinotta, F. A.; Wilson, J. W.; Gladman, B; Horneck, G; Lindegren, L.; Melosh, J.; Rickman, H.; Valtonen, M.; Zheng, J. Q. (2000) Natural transfer of viable microbes in space. 1. From Mars to Earth and Earth to Mars. *Icarus* **145,** 391 - 428.

44. Odum, H. T., and Hoskins, C. M. (1957) Metabolism of a laboratory stream microcosm. *Publ. Inst. Mar. Sci. Univ. Texas* 4, 115-133.

45. O'Leary, B. T. (1977) Mining the Apollo and Amor asteroids. *Science* 197, 363-364.

46. O'Neill, G. K. (1974) The colonization of space. *Physics Today* 27, 32-38.

47. Oro, J., Mills, T. and Lazcano, A. (1995) Comets and life in the universe. *Adv. Space Res.* 15, 81-90.

48. Shearer, C. K., Papike, J. J. and Rietmeijer, F. J. M. (1998) The planetary sample suit and environments of origin. In *Planetary Materials* (J. J. Papike, ed.) Reviews in Mineralogy v. 36, Mineralogical Society of America, Washington, pp. 1-1 – 1-28.

49. Steele, A., K.; Thomas-Kerpta, F. W.; Westall, R.; Avci, E. K.; Gibson, C.; Griffin, C.; Whitby, C.; McKay, D. S.; and Toporski, J. K. W. (2000) The microbiological contamination of meteorites: A null hypothesis. *First Astrobiology Science Conference.* NASA Ames Research Center, Moffett Field, California, April 2000. p. 23.

50. Stoker, C. R., Boston, P. J., Mancinelli, R. R., Segal, W., Khare, B. N., and Sagan, C. (1990) Microbial metabolism of tholin. *Icarus* 85, 241-256.

51. Tomeoka, K. and Buseck, P. R. (1985) Indicators of aqueous alteration in CM carbonaceous chondrites: Microtextures of a layered mineral containing Fe, S, O and Ni. *Geochim. Cosmochim. Acta* 49, 2149-2163.
52. Urey, H. C. (1962). Lifelike forms in meteorites. *Science* **137**, 625-628.
53. van Veldhoven, P. P. and Mannerts, G. P. (1987) Inorganic and organic phosphate measurements in the nanomolar range. *Anal. Biochem.* 161, 45-48.
54. Winogradsky, S. (1949) *Microbilogie du sol, problems et methodes.* Barnoud Freres, France.

Table 1. Solute concentrations (mg/l) in extracts of the Allende and Murchison meteorites, obtained by extractions at low and high solid/solution ratios.

	Ca[a]	Mg[a]	Na[a]	K[a]	Mn[b]	Fe[b]
Allende ($r_{s/w}$ = 0.02 g/ml)	1.8	2.0	1.3	0.44	0.014	0.008
Allende ($r_{s/w}$ = 10 g/ml)	420	380	340	180	7	4
Murchison ($r_{s/w}$ = 0.02 g/ml)	64	56	52	4.2	0.22	0.18
Murchison ($r_{s/w}$ = 10 g/ml)	14,000	17,000	19,000	1,100	110	90
soil solutions, median and range[e]	32 (1-60)	25 (0.7-100)	15 (9-30)	3.5 (1-11)	(0.02-2)	(0.1-0.25)
bacterial biomass ($\mu g/g$)[f]	5,100	7,000	4,600	1.2E5	30	250
c (CFU/ml) Allende[g]	4.1E10	2.7E10	3.7E10	7.5E8	1.2E11	8.0E9
c (CFU/ml) Murchison[g]	1.4E12	1.2E12	2.1E12	4.5E9	1.8E12	1.8E11

Table 1 (continued)

	Ni^b	Cl^a	$NO_3\text{-}N^a$	$SO_4\text{-}S^a$	$PO_4\text{-}P^c$	C^d (organic)
Allende ($r_{s/w}$ = 0.02 g/ml)	1.7	0.7	0.044	1.2	0.08	2
Allende ($r_{s/w}$ = 10 g/ml)	850	730	68	320	44	1,000
Murchison ($r_{s/w}$ = 0.02 g/ml)	2.0	7.6	0.22	188	0.1	20
Murchison ($r_{s/w}$ = 10 g/ml)	1,000	1,800	120	57,000	51	10,000
soil solutions, median and range[e]		10 (7-50)	(2-800)	5 (3-5000)	0.005 (0.001-30)	
bacterial biomass ($\mu g/g$)[f]		2,300	9.6E4	5,300	3.0E4	4.9E5
c (CFU/ml) Allende[g]		1.6E11	3.5E8	3.0E10	7.3E8	1.0E9
c (CFU/ml) Murchison[g]		3.9E11	6.3E8	5.4E12	8.5E8	1.0E10

Footnotes to Table 1

a. Concentrations (mg/l or ppm) in extracts obtained at 20 °C for 4 days at low (average data from 0.02-0.1 g/ml) and high (1-10 g/ml) solid/solution ratios. The corresponding concentrations of extractable elements in the solids (c_{solid} mg/g) may be obtained by dividing the listed values by $10^3 r_{solid/solution}$ (g/ml). The average values obtained at solid/solution ratios of 1, 2, 4 and 10 g/ml are given for the 10 g/ml extractions. Estimated uncertainty ±30%, except Cl and NO_3-N measured at low concentrations in extracts obtained at $r_{solid/solution}$ = 0.1 - 1.0, where an uncertainty by of a factor of 2 may apply.

b. Based on c_{solid} measured by extraction by 1M NH_4OAc for 24 hours at a solid/solution ratio of 0.028 g/ml. Solution concentrations calculated using this c_{solid} value and $r_{solid/solution}$ = 0.02 or 10 in equation (1).

218

c. Based on c_{solid} measured by extraction in 4 days at a solid/solution ratio of 1.0 g/ml (Mautner and Sinai, 2001), and using this c_{solid} value and $r_{solid/solution} = 10$ and in equation (1).

d. Estimated as half of the yield of organic carbon obtained at 121 °C for 15 minutes at solid/solution ratios of 0.01 - 0.04 (Mautner et al., 1995).

e. Elemental concentrations in soil solution. Concentrations in ppm (Bowen, 1966).

f. Elemental concentrations in bacterial dry biomass in ppm (Bowen, 1966).

g. Calculated maximum bacterial populations (CFU/ml) allowed by the concentration of a given nutrient in the extracts of Allende and Murchison obtained by extractions at $r_{solid/solution} = 10$ g/ml as given in rows 2 and 4. Calculated using equation (4) for bacteria with radius of 1 μm, dry mass of 2×10^{-12} g, and elemental content per gram bacterial dry mass as in row 6 (Bowen, 1966).

Table 2. Microorganisms and populations observed in meteorite cultures inoculated with a mix microbial population isolated from a peat bog extract.

Biolog 96-well carbon source ID	Gram Stain, catalase and oxydase tests	Biolog sim and distance[a]	Allende extract[b]	Allende wet solid[c]
Clavibacter michiganense	GP rod, Cat +, Oxy +	0.398, 1.74	8E4	1.0E6[c] 2.8E6[d]
Microbacterium imperiale	GP rod, Cat +, Oxy +	0.278, 12.78	6E4	4.6E5[c] 4.0E5[d]
Eureobacterium saperdae	GP rod, Cat +, Oxy +	0.706, 4.35	6E4	-
Pseudomonas putida	GN, Oxy +	0.613, 5.99	6E4	1.6E5[c] 4.0E4[d]
Corynebacterium sp.	GP rod, Oxy -		1.6E6	4.2E5[c] 3.0E4[d]
yeast			1E4	
filamentous fungus			1E4	
Total population			1.9E6	2.0E6[c] 4.7E6[d]

Table 2 (continued)

Biolog 96-well carbon source ID	Murchison extracts[b]	Murchison wet solid[c]	Sand solution[b]
Clavibacter michiganense	6.6E5	9.0E5[c] 2.0E6[d]	1.6E6
Microbacterium imperiale	3.4E5	9.4E5[c] 3.4E5[d]	-
Eureobacterium saperdae	-	-	-
Pseudomonas putida	1E5	-	1.2E5
Corynebacterium sp.	3E6	3.7E6[c] 2.8E4[d]	
yeast			
filamentous fungus	7E4	2.6E5	2E5
Total population	4.2E6	5.8E6[c] 4.8E6[d]	1.9E6

a. Indicators of reliability of Biolog plate identification.
b. Post log phase steady-state populations in meteorite and sand extracts cultured for 31 days. Extracts refer to cultures in extracts obtained at solid/solution ratios of 0.02 g/ml.
c. As in footnote b, for populations in the liquid phase over the wetted solids after 31 days. Wet solids refer to cultures at solid/solution ratio of 1 g/ml.
d. As in c, populations in the same cultures after 8 months. In addition to the species listed, the long-term Allende cultures contained a new species with globular yellow colonies, 1.2E6 CFU/ml and with small green-yellow colonies, 2.1E5 CFU/ml. The long-term Murchison cultures contained the species with globular yellow colonies, 1.2E6 CFU/ml and a species with flat cream yellow colonies, 1.1E6 CFU/ml.

Table 3. Algal populations in meteorite extracts and wet solids (CFU/ml)[a]

	green unicell-ular[b]	brown diatom[b]	blue-green filamentous	total
Initial, all extracts	1E4	2E2	8E1	1.0E4
Allende	4.8E4	4E3	1E3	5.3E4
Murchison	4E4	8E3	4E2	4.8E4
Sand	4E4	1E4	4E2	4.8E4
wet solid				
initial, all solids	5E5	1E4	4E3	5.1E5
Allende	0	4E4	0	4E4
Murchison	4E5	1E4	2E3	4.1E5
Sand	8E4	8E4	2E3	1.6E5

a. Algal populations in the mixed bacterial and algal cultures, after the initial inoculation and after 15 days. Extracts refer to cultures in extracts obtained at solid/solution ratios of 0.02 g/ml. Wet solids refer to cultures at solid/solution ratio of 1 g/ml. b. Green unicellular algae - *Chlorella sp.* and *Chlorosarcinopsis sp.*; Brown diatom - *Navicula sp.;* - Filamentouos blue-green *Leptolyngbya sp.* and *Phormidium sp.*

Figure Legends

Figure 1. Concentrations of cations in aqueous solutions, c_{aq} (mg/l) in extracts of the Allende meteorite at various solid/solution ratios (g/ml).
Figure 2. Microbial populations in Allende extracts. Cm - *Clavibacter michganenese*; Mi - *Microbacterium imperiale*; Es - *Eureoubacterium saperdae*; Pp - *Pseudomonas putida*; Cs - *Corynebacterium sp.*
Figure 3. Microbial populations in Murchison extracts. Notation as in Figure 2.
Figure 4. Microbial populations in extracts of acid-washed sand. Notation as in Figure 2.

Appendix A3.2

Icarus 2002, 158, 72-86

PLANETARY BIORESOURCES AND ASTROECOLOGY. 1. PLANETARY MICROCOSM BIOASSAYS OF MARTIAN AND CARBONACEOUS CHONDRITE MATERIALS: NUTRIENTS, ELECTROLYTE SOLUTIONS, AND ALGAL AND PLANT RESPONSES

Michael N. Mautner, Soil, Plant and Ecological Sciences Division, Lincoln University, Lincoln, New Zealand, and Department of Chemistry, University of Canterbury, Christchurch 8002, New Zealand (mautnerm@lincoln.ac.nz)

Abstract

The biological fertilities of planetary materials can be assessed using microcosms based on meteorites. This study applies microcosm tests to Martian meteorites and analogues, and to carbonaceous chondrites. The biological fertilities of these materials are rated based on the soluble electrolyte nutrients, the growth of mesophile and cold-tolerant algae, and plant tissue cultures. The results show that the meteorites, in particular the Murchison CM2 carbonaceous chondrite and DaG 476 Martian shergottite contain high levels of water-extractable Ca, Mg and SO_4-S. The Martian meteorites DaG 476 and EETA 79001 also contain higher levels of extractable essential nutrients NO_3-N (0.013-0.017 g kg^{-1}) and PO_4-P (0.019-0.046 g kg^{-1}) than the terrestrial analogues. The yields of most of the water-extractable electrolytes from the meteorites and analogues vary only by factors of 2 - 3 under a wide range of planetary conditions. However, the long-term extractable phosphate increases significantly under a CO_2 atmosphere. The biological yields of algae and plant tissue cultures correlate with extractable NO_3-N and PO_4-P, identifying these as the limiting nutrients. Mesophilic algae and *Asparagus officinalis* cultures are identified as useful bioassay agents. A fertility ranking system based on microcosm tests is proposed. The results rate the fertilities in the order Martian basalts > terrestrial basalt, agricultural soil > carbonaceous chondrites, lava ash > cumulate igneous rock. Based on the extractable materials in Murchison, concentrated internal solutions in carbonaceous asteroids (3.8 mol L^{-1} electrolytes and 0.1 mol L^{-1} organics) can support microorganisms in the early Solar System. The bioassay results also suggest that carbonaceous asteroids and Martian basalts can serve as soils for space-based agriculture. Planetary microcosms can be applied in experimental astroecology for targeting astrobiology exploration and for identifying space bioresources.

Introduction

Rocks and soils in early aqueous planetary environments may have provided resources for the origins of life and nutrients for early microorganisms. The objects of interest include carbonaceous asteroids during aqueous alteration, and local igneous rocks or comets and meteorites that land aqueous planetary environments. Similar materials are also of interest as potential soils for space-based agriculture (Ming and Henninger 1989). The present work demonstrates the use of planetary microcosms based on meteorites and terrestrial analogues to assay the biological potentials of these materials.

Meteorites can play several roles in which biological fertility is significant. For example, carbonaceous chondrites and similar dust or cometary materials import organics to planets (Chyba and Sagan 1992, Greenberg and Li 1998). Carbonaceous chondrite asteroids during aqueous alteration and carbonaceous meteorites on aqueous planets contain water in their pores and may form highly concentrated solutions that facilitate biogenesis (Mautner et al. 1995; Mautner 1997a; Bodnar and Zolensky 2000; Cohen and Coker 2000). Meteorites may also actively transport microorganisms amongst asteroids and planets (Arrhenius 1908, Crick and Orgel 1973, Chyba and McDonald 1995, Mautner 1997a; Mileikowsky *et al.*, 2000). After impact on aqueous planets, the meteorites will constitute the first nutrient environments for the embedded microorganisms (Mautner 1997a, Mautner *et al* 1997).

There are indications that many space materials can indeed support life. First, their mineral and organic constituents are similar to terrestrial rocks that support diverse geo-microbiology. Algae growing on meteorite dust in Greenland were observed as early as 1870 (Leslie, 1879, Maurette *et al.* 1986). The nutrient values of organic planetary materials were demonstrated on a synthetic terrestrial analogue, tholin (Stoker *et al.* 1990). The Murchison CM2 meteorite was observed to have soil fertility parameters comparable to productive terrestrial soils (Mautner 1997b, Mautner 1999). Murchison extracts were observed to support various soil microorganisms such as the oligotrophs *Flavobacterium oryzihabitans* and *Nocardia asteroides*. Experiments with *Pseudomonas fluorescence* showed that meteorite organics can serve as the sole carbon source (Mautner *et al.* 1997). Indications were also found that the Murchison materials support the anaerobic thermophile eubacterium *Thermotoga maritima* and the aerobic thermophile *Thermus aquaticus* (H. W. Morgan, quoted in Mautner *et al.* 1997). In contrast, the Allende meteorite was observed to inhibit biological growth. Martian minerals may have also supported microorganisms in the past (McKay *et al.* 1996). Various carbonaceous and Martian meteorites were found recently to contain microorganisms from terrestrial contamination (Steele *et al.* 2000).

These observations suggest that diverse planetary materials can support biological activity. Their biological properties, as represented by meteorites or future return samples, should be assayed systematically in a way similar to agricultural soils (McLaren and Cameron 1996, Beare *et al.* 1997). Previous microbial experiments with meteorites (Mautner 1997a) constituted limited planetary microcosms for such purposes.

The objectives of the present study are: (1) to introduce planetary microcosms as tools for the bioassay of planetary materials; (2) to measure the contents of water-extractable nutrient electrolytes in meteorites and analogues; (3) to use the extracts in microcosms for testing biological responses; (4) to identify useful bioassay agents; (5) to check the consistency of various nutrient and biological assays, and to develop a rating method based on the assays; (6) to apply microcosm rating to the biological fertilities of some actual and simulated planetary materials.

Experimental

1. Materials

The rock samples used were Martian and carbonaceous chondrite meteorites and terrestrial analogues. The Dar al Gani 476 (DaG 476) meteorite and Elephant Morraine 79001 (EETA 79001) lithology A are both basaltic shergottites (McSween and Jarosewitz 1983). DaG 476 contains a fined-grained pyroxene and feldspathic glass ground-mass containing also sulphides and phosphates. Both meteorites contain olivine, orthopyroxene and chromite. The DaG 476 meteorite was subject to extensive terrestrial weathering, leading to the formation of carbonates. Phosphate minerals in the shergottites include merrilite and chlorapatite.

The comparative mineralogy of DaG 476 and EETA 79001 was discussed recently (Zipfel *et al.* 2000).

The Nakhla meteorite is a cumulate igneous rock. Its main component is augite, with some olivine and minor other minerals (McSween and Treiman 1998). As only small amounts of Nakhla were available, the Theo's Flow lava formation in Canada that was described as closely similar in mineralogy was used as a terrestrial analogue (Friedman 1998). A further basalt sample containing 65% labradorite feldspar, 25% clinopyroxenite and 10% magnetite from Timaru, New Zealand was also used.

Further terrestrial analogues included NASA simulants of lunar and martian soils. The Mars soil simulant JSC Mars-1 is a sample of lava ash from the Pu'u Nene volcano in Hawaii (Allen *et al.* 1998). It contains Ca-feldspar and minor magnetite, olivine, augite pyroxene and glass, including a highly weathered glassy matrix. It also contains nanophase ferric oxide similar to that inferred for Martian soil. The lunar simulant JSC-1 is a glass-rich volcanic ash from the San Francisco volcanic field near Flagstaff, Arizona (McKay *et al.* 1993). The elemental composition is similar to Apollo 14 soil sample 14163, and contains plagioclase, clinopyroxene, orthopyroxene, olivine, magnetite, ilmenite and apatite. In addition, a representative terrestrial agricultural soil, from the Templeton area in New Zealand, Udis Ustochrept, fine loamy mixed, mesic soil, was also used for comparison.

Two carbonaceous chondrites, Allende and Murchison, were also used. The mineralogies of both are well known (Fuchs *et al.* 1973, Bunch and Chang 1980, Komacki and Wood 1984) and were reviewed recently (Brierly and Jones 1998). The main component of Murchison is a phyllosilicate formed by aqueous alteration in the parent body.

2. Extraction and Analysis

Samples of meteorites and terrestrial analogues were ground in agate mortar to yield particle size distributions similar to terrestrial soils (Mautner and Sinaj 2002). Extractions were carried out in polythene tubes washed in 10% acetic acid for 24 hours and rinsed four times with deionized water to remove electrolyte impurities.

The extractions were performed using deionized and Millipore filtered water with resistivity >18 Mohm-cm. The water was degassed by bubbling with N_2, denoted as "H_2O/N_2" (Table 1 below). In some extractions, the effects of early planetary CO_2 atmospheres were simulated by extracting the samples in deionized water saturated with CO_2 at a pH of

3.9, with the extraction vials sealed under CO_2 and placed in sealed jars filled with CO_2 as a further barrier against exchange with the atmosphere. These conditions are denoted as "H_2O/CO_2" below.

The powders were extracted at solid/solution ratios and extraction times as described in Table 1a and 1b. Extractions at 20°C were carried out with shaking on an orbital shaker. Hydrothermal extracts were obtained by placing the sealed extraction tubes in an autoclave and extracting under standard sterilizing conditions at 121 °C for 15 minutes. For analysis, the powders were separated from the solution by centrifuging.

Trace metals were extracted from Allende and Murchison at $r_{solid/solution} = 0.027$ kg L^{-1} by 1M NH_4OAc solution, a standard soil extractant (Blakemore et al., 1987), and analyzed by ICP-MS at Hill Laboratories, Hamilton, New Zealand.

Anions in the extracts were analyzed by ion exchange chromatography using a Waters Ion Exchange Chromatograph and Waters Baseline 810 software. The method used was Waters Ion Chromatography Method A-102 "Anion Exchange Analysis Using IC-Pak A HC Column Borate/Gluconate Eluent", with the samples filtered through a 0.1 micron filter paper prior to analysis. Cations were analyzed by a Shimadzu AA-6200 Atomic Absorption Flame Emission Spectrophotometer.

Table 1a.   Concentrations of water-extractable cations (g kg^{-1}) in carbonaceous chondrites, Martian meteorites, Martian and Lunar soil analogues and terrestrial soil solutions.

Materials	Extraction Conditions[a]	Ca	Mg	Na	K
Carbonaceous Chondrites					
Allende	HT[b]	0.097	0.20	0.060	0.034
	H_2O/CO_2[d]	0.20	0.30	0.30	0.034
Murchison	HT[b]	3.0	4.0	1.4	0.34
	H_2O/CO_2[d]	2.8	1.7	2.4	0.18
Mars Meteorites					
Dar al Gani 476	HT[b]	1.0	0.38	0.067	0.064
	H_2O/CO_2[d]	1.1	0.58	0.040	0.032
EETA 79001	HT[b]	0.18	0.084	0.076	0.016
	H_2O/CO_2[d]	0.42	0.17	0.043	0.006
Mars and Lunar Analogues					
Basalt	HT[b]	0.09	0.037	0.088	0.032
Theo's Flow	HT[b]	0.24	0.077	0.034	0.007
	H_2O/CO_2[d]	0.81	0.14	0.026	0.012
JSC Mars Simulant	HT[b]	0.15	0.014	0.011	0.11
	H_2O/CO_2[d]	0.36	0.08	0.027	0.13
Lunar simulant	HT[b]	0.16	0.004	0.10	0.027
Terrestrial Soils					
Templeton soil	HT[b]	0.04	0.004	0.04	0.03
Terrestrial Soil		0.001-0.060	0.0007-0.1	0.009-0.03	0.001-0.011

Materials	Extraction Conditions[a]	Ca	Mg	Na	K
Solution[I]					

Table 1b. Concentrations of water-extractable anions (g kg^{-1}) in carbonaceous chondrites, Martian meteorites, Martian and Lunar soil analogues and terrestrial soil solutions.

Materials	Extraction conditions[a]	Cl	NO$_3$-N	SO$_4$-S	PO$_4$-P
Carbonaceous Chondrites					
Allende	HT[b]	0.10	0.004	0.36	0.0075[c]
	H$_2$O/CO$_2$[d]	0.06	0.003	0.38	0.005
Murchison	HT[b]	0.44	0.008	7.6	0.0050[c]
	H$_2$O/CO$_2$[d]	0.28	0.008	6.8	0.0048
Mars Meteorites					
Dar al Gani 476	HT[b]	0.074	0.017	0.92	0.019
	H$_2$O/CO$_2$[d]	0.06	0.015	0.88	0.024[f]
EETA 79001	HT[b]	0.037	0.013	0.048	0.046
	H$_2$O/CO$_2$[d]	0.24	0.013	0.031	0.037[g]
Mars and Lunar Analogues					
Basalt	HT[b]	0.056	0.002	0.008	0.013
Theo's Flow	HT[b]	0.044	0.002	0.008	0.001
	H$_2$O/CO$_2$[d]	0.03	<0.001	0.004	0.0002
JSC Mars Simulant	HT[b]	0.048	0.004	0.012	0.003
	H$_2$O/CO$_2$[d]	0.02	0.002	<0.001	0.0006
Lunar Simulant	HT[b]	0.048	0.002	0.018	0.017
Terrestrial Soils					
Templeton Soil	HT[b]	0.018	<0.001	0.007	0.001[h]

Materials	Extraction conditions[a]	Cl	NO$_3$-N	SO$_4$-S	PO$_4$-P
Soil Solution[I]		0.007 -0.05	0.002- 0.80	<0.003- 5	0.000001- 0.03

Footnotes to Table 1a and 1b.

a. All extractions at solid/solution ratios of 0.1 g mL^{-1} unless noted otherwise.

b. Hydrothermal extractions at 121 °C for 15 minutes. Average results of four replicates with standard deviations as shown in parentheses.

c. Average of present measurements (Allende 0.0051 g kg^{-1}, Murchison 0.0045 g kg^{-1}) and literature results (Allende 0.0098 g kg^{-1}, Murchison 0.0055 g kg^{-1}) (Mautner 1997a).

d. Extractions in water saturated with carbon dioxide (H_2O/CO_2) at a solid/solution ratio of 0.1 kg L^{-1} at 20 °C for 24 hours.

e. Extraction in H_2O/N_2 at 20°C for 4 days. Solid/solution ratio as noted; >1 kg L^{-1} represents the average of approximately constant extractable amounts at solid/solution ratios of 1, 2, 4 and 10 kg L^{-1}.

f. Extraction in H_2O/CO_2 at a solid/solution ratio of 0.01 kg L^{-1} at 20° C for 8 days.

g. Extraction in H_2O/CO_2 at a solid/solution ratio of 0.1 kg L^{-1} at 20° C for 1 day.

h. Extraction in H_2O/N_2 at a solid/solution ratio of 0.1 kg L^{-1} at 20° C for 4 days.

i. Soil solution values units of g L^{-1} (Bowen 1966).

Analysis of the extracted phosphate was performed by developing the solutions with malachite green reagent at 4:1 solution/reagent ratios for one hour and measuring the absorbance at 630 nm (van Veldhoven and Mannaerts 1987). In addition, total phosphorus content was determined by extraction of 10 - 100 mg samples with 10 ml of 70% HNO_3 solution overnight, adding 4 ml perchloric acid and heating to 210 °C, cooling and diluting appropriately for analysis by the malachite green reagent. Isotope exchange kinetics (IEK) assays of bioavailable phosphate were performed on a suspension of 100 mg solid in 10 ml deionized water that was allowed to equilibrate with the solution for 24 hours. A small amount of $^{32}PO_4^{3-}$ was then added and solution samples were withdrawn 1, 10, 30 and 60 minutes after introducing the labelled tracer and the radioactivity in the solution was measured in a scintillation counter (Frossard and Sinaj 1997).

The meteorite sample sizes of 10 – 40 mg that are available per analysis are much smaller than the 1 – 10 gram samples used in standard soil analysis. The small samples required soil microanalysis methods that were developed for meteorites previously (Mautner 1997a). The microanalysis methods for one of the analytical procedures, extractions with 1M ammonium acetate, could be compared with conventional analysis by 10 -30 laboratories worldwide, compiled by the International Soil Analytical Exchange Program (Wageningen University, Dept. of Environmental Sciences, The Netherlands, WEPAL@MAIL.BenP.wag-ur.NL). The microanalysis data fell in the reported range. Comparison with the SAEP results leads to an estimated coefficient of variation (cv = std/mean) of 0.3 in our microanalysis measurements. This uncertainty estimate is consistent with the average cv of the results in Table 1 obtained from four replicate samples in each extraction. In addition to the usual factors in the SAEP inter-laboratory scatter, the uncertainties in our data are due to the small available sample sizes. This leads to low concentrations of the analysed elements in the extracts, often close to the detection limits of 10^{-4} - 10^{-3} g L^{-1} for anions and cations and 10^{-5} g L^{-1} for phosphate.

3. Plant and Algal Cultures

For the plant tissue culture experiments, the meteorites and analogues were extracted and sterilised at 121 °C for 15 minutes at the ratio of 100 mg solid powder in 0.5 ml deionized water, except EETA 79001 where 10 mg solid was extracted in 0.2 ml water. For each plant culture medium, 45 μL of the extracts and 5 mg of the extracted solid powder were transferred to a 2 ml polythene microfuge tube. For blanks, deionized water was used instead of the extracts. Arabidopsis seeds were germinated and grown in these extracts directly. For asparagus and potato tissue cultures, to each 45 μL of extracts or water, a further 45 μL of 5 mM NH_4NO_3 solution + 3% sucrose was added as nitrogen and carbon sources, leading to 2.5 mM NH_4NO_3 and 1.5% sucrose in the culture solutions.

Algal culture media were prepared by extracting 20 mg of the solids in 0.2 ml deionized water and mixing with equal volumes of 10 mM NH_4NO_3 or with deionized water. Two culture methods were used. In the first method, for testing benthic growth, both the extract and the extracted solid powder were placed in a cavity microscope slide partially covered with a slide cover, and kept in a closed dessicator at 100 percent humidity to prevent evaporation. This somewhat reduced the light flux and these cultures were grown <80 μE m^{-2} s^{-1}. In the second method, used for population counting experiments, the aqueous extracts and only traces of the extracted powder were used under full illumination at 80 μE m^{-2} s^{-1} in sealed vials which were opened periodically for air exchange.

The inoculant algal cultures were grown in standard nutrient medium, and washed by four cycles of centrifuging and washing with deionized water. This procedure dilutes the nutrient medium remaining in the algal sample by factors of 10^6 - 10^8, to negligible levels. The final washed algal pellets were dispersed in deionized water and diluted to give inoculations used 20 μL of the mixed and washed algal cultures to yield starting populations of 10^4 CFU ml^{-1} of each algae in the microcosms. Algal populations in 20 μL samples were counted by direct microscopic count in a haemocytometer chamber.

Plant tissue cultures were established from *in vitro Asparagus officinalis*, cultivar "Limbras 10" genotype ASC 69. Apical meristem shoot tips about 1 mm long were dissected from 4 – 8 cm plants grown on agar. The parent plants developed to full size in agar in one month but were kept in agar for an additional three months to assure that all the nutrients in the medium were exhausted. This procedure produces plants that are depleted of nutrients, making them more responsive to nutrients in the mineral extracts. Well formed globular apical shoot tips could be removed uniformly from each plant. Potato plants *Solanum tuberosum* cultivar Iwa were grown using a similar procedure. Plants of *Arabidopsis thaliana* strain "Landsberg Erecta" were germinated from seed on filter paper wetted with deionized water and grown to about 10 mm full size that is achievable on the seed nutrients alone, before introduction to the culture media.

All plant cultures were grown for 20 days in closed microfuge tubes in standard tissue culture growth chambers under illumination by cool white fluorescent lights with an incident light flux on the samples of 80 μE m^{-2} s^{-1} using 16 h light/8 h dark photocycles. Algal cultures were grown in similar growth chambers in vials with a punctured cap that allowed air exchange. The algal cultures were contained in glass jars saturated with water vapor to allow full light exposure but prevent evaporation.

Results

1. Extractable nutrient electrolytes in Mars meteorites and analogues

The concentrations of extractable materials in the powdered solids are listed in Table 1. The majority of the reported values are the average of four replicate measurements with the standard deviations as listed. The average of the cv values (std/mean) for these measurements is 0.28, consistent with our general experience in similar extraction measurements. This may be assumed as the estimated uncertainty also for those results in Table 1 where less replicates were performed due to the scarcity of materials.

The water-extractable solutes may be calculated in terms of the concentration of extractable element in the solid, c_{solid} (g kg^{-1}), and the concentration established in the extract solution c_{aq} (g L^{-1}). For a given extraction, the two quantities are related by equation (1).

$$c_{aq} \text{ (g L}^{-1}) = c_{solid} \text{ (g kg}^{-1}) \, w_{solid} \text{ (kg)} / V_{aq} \text{ (L)} =$$
$$c_{solid} \text{ (g kg}^{-1}) \, r_{solid/solution} \text{ (kg L}^{-1}) \qquad (1)$$

Here w_{solid} is the weight of solid subject to extraction, V_{aq} the volume of the extract and $r_{solid/solution}$ is the solid/solution ratio.

2. Anions and Cations

The amounts of extracted elements depend on several variables such as the solid/solution ratio, pH and temperature. Most of the data in Table 1 relate to extraction under mild hydrothermal conditions (121 °C, 15 minutes and at a solid/solution ratio of 0.1 kg L^{-1}). Extractions under more moderate conditions at 20°C in H_2O/N_2 generally yielded comparable amounts usually lower up to a factor of two, as can be observed by comparing in Table 1 in the Murchison results at 0.1 kg L^{-1} for hydrothermal vs. H_2O/N_2 extractions. Using higher solid/solution ratios generally decreases the extracted amounts per unit weight of the solid as also observed in Table 1 for Murchison, but the effect is small over the wide range of solid/solution ratios shown. Varying the extraction times in Table 1 from 2 – 8 days and in some cases up to 30 days changed the extracted amounts only within a factor of two in most cases.

Table 1 also shows the effects of extraction under a 1 atm. CO_2 atmosphere, which models the atmospheres of early Earth and Mars (Kasting and Mishna 2000). The resulting high levels of dissolved carbonic acid and low pH can affect the bioavailabilities of ions. Table 1 shows that the amounts extracted hydrothermally in H_2O/N_2 or at 20°C by H_2O/CO_2 are generally similar, changing the extracted amounts by less than a factor of two in most cases.

Altogether, the extractions in Table 1 cover a wide range of planetary and asteroid conditions, from 20 to 121 °C and, for Murchison, from an excess of water to an excess of solid and also acidic solutions (pH

3.9) under CO_2 dominated atmospheres. The results show that the amounts of extractable electrolytes vary only moderately, generally less than a factor of two, under a wide range of plausible planetary conditions.

The nutrients extracted from the meteorites may be compared in Table 1 with those extracted from terrestrial analogues. The highest levels of extractable Ca, Mg Na, K, Cl and SO_4 were found in Murchison, up to an order of magnitude higher than in most of the other materials.

Although quite different in origin and mineralogy, the Martian DaG 476 also yielded higher Ca, Mg and SO_4-S than the other materials, and K and Cl also in the high range. On the other hand, the other Martian sample, EETA 79001, yielded soluble electrolytes comparable with the terrestrial samples. Notable in the Martian meteorites are the relatively high levels of the limiting nutrients NO_3-N and PO_4-P compared with in the other materials.

The 7.6 mg g^{-1} sulphate extractable from Murchison was about two orders of magnitude higher than in the terrestrial materials. The high levels of Ca and SO_4-S may be derived from gypsum crystals observed on the Murchison surfaces (Fuchs *et al.* 1973) and may also be related to the high total S content of 3% in Murchison (Jarosewich 1971, Lovering *et al.* 1971)). Note that the extractable sulphate is significantly higher in all the meteorites than in the terrestrial materials examined. The lowest amounts of most extractable electrolytes were found in the sample terrestrial soil which, unlike the meteorites and rocks, have been subject to extensive weathering and leaching, although even with these amounts of nutrients, it is a productive agricultural soil.

In most of the electrolytes, the Allende CV3 meteorite yielded less soluble amounts by factors of 5 - 40 lower than Murchison, but comparable to the terrestrial rocks, except for relatively high Mg and SO_4-S. A few extractions, however, yielded several times less Ca, Mg and SO_4-S than the results in Table 1, suggesting that Allende may be inhomogenous in these components on the scale of 20 - 100 mg samples.

The concentrations in the extracts (c_{aq}, g L^{-1}) can be obtained according to equation (1) by multiplying the c_{solid} values in Table 1 by the solid/solution ratio, which is 0.1 g kg^{-1} in most samples. The resulting concentrations can be compared with those in terrestrial soil solutions shown in Table 1. These solutions, diluted by a factor of two, were applied in the algal and plant cultures below. Except for the large concentrations obtained from Murchison, the other meteorites and simulants yielded solution concentrations within the range of terrestrial soil solutions. However, the limiting nutrient NO_3-N was lower by about an order of magnitude in most of the extracts by about an order of magnitude than the

234

lower limit of soil solutions, even for the Martian meteorites which were the richest in extractable nitrate.

The other limiting nutrient is extractable phosphate. Because of its significance, it is desirable to characterize available phosphate in more depth. A quantitative assessment of long-term availability can be obtained by the isotope exchange kinetic method (Frossard and Sinaj 1997). The method, as described above in the Experimental section, is suitable for meteorite studies as it can use small, 100 mg samples.

This method can assess, for example, phosphate available through extraction by a root system during three months. Applied to the present materials, isotope exchange measurements showed that the phosphate available in three months is higher by factors of 17-200 for the materials used than the short-term soluble amount. An exception was DaG 476 where this ratio was only 1.6, possibly because the amount of soluble P is already high. The amount of (PO_4-P) (3 months) was higher by about an order of magnitude in the carbonaceous chondrites than in the other materials. Altogether, 13.0% and 43.9% of the total elemental P content was long-term bioavalable in Allende and Murchison while only 1 - 6 % of the P present was found to be bioavailable in the igneous materials.

The effects of extraction by H_2O/CO_2 for 21 days on the long-term available PO_4-P was substantial. This protracted exposure to H_2O/CO_2 increased the available P from 13.0 to 51.6% of the total P content in Allende, from 43.9 to 90.4% in Murchison, from 1.6 to 11.6% in DaG 476, from 6.4 to 97.6% in Theo's Flow, and from 2.5 to 40.5% in Hawaii lava. A more extensive study of extractable and of long-term bioavalable phosphate in planetary materials will be presented elsewhere (Mautner and Sinaj 2002).

3. Plant Bioassays

In previous studies, asparagus and potato tissue cultures showed that Murchison had a nutrient effect, with optimal growth from extractions at solid/solution ratios of 0.05 – 0.1 kg L^{-1}. In contrast, the extracts of Allende indicated an inhibitory, possibly toxic effect (Mautner 1997a, Mautner et al. 1997).

The present work examined the effects of extracts of further planetary materials. The concentrations of the nutrients in the various extracts may be calculated from Table 1 as discussed above. Two sets of each culture were grown with the combined number of replicate plants as shown in Table 2. Significant differences were observed amongst the various media. The products in the control water blanks, supplemented only by nitrate and sucrose, showed the least growth, possibly only cell enlargement supported by stored nutrients in the starting meristem tips. In comparison, the meteorite and lava ash products showed more coloration and development. The plants in the extracts of the Hawaii lava Mars simulant and some of the DaG 476 products showed more stem development than the Murchison products. Some of the Murchison products showed a partial reddening that may indicate phosphorus deficiency. The DaG 476 products showed the most differentiated development and the deepest green coloration, possibly due to the high phosphate content.

Cultures in one set of the Allende extracts showed brown coloration, decreased size and lower fresh weights even than the water blank, as was observed also previously (Mautner 1997a). These effects suggest low nutrient concentrations or the presence of toxic elements. However, extracts from a different fragment of Allende showed more green coloration and development, and one of the highest yields. The differences between the two sets may reflect the inhomogeneity noted above in the concentrations of extractable elements from different Allende fragments. The inhibitory fragment is denoted in Table 2 as Allende A and the nutrient fragment as Allende B

Table 2 shows the yields determined as fresh weights, since the dry masses of <0.1 mg were too small to measure. Statistical analysis was performed to calculate the Mann-Whitney p values comparing the weights of the product plants in each pair of media. The p value obtained is inversely related to the statistical difference, and a value of $p < 0.050$ indicates a statistically significant difference between the two sets. A stricter but less sensitive indicator may be obtained using protected mean separation tests.

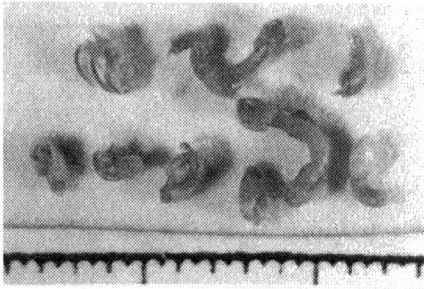

Dag 476

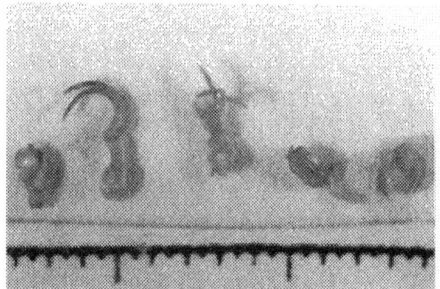

Murchison

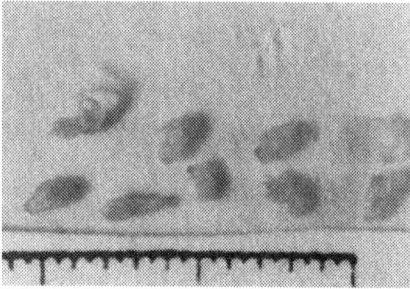

Water

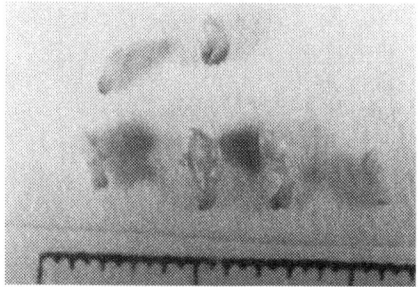

Hawaii lava Mars simulant

FIG. 1. Plant tissue cultures of *Asparagus officinalis* in meteorite and soil extracts, all supplemented with 5 mML^{-1} NH$_4$NO$_3$ and 3% sucrose. Small ticks are 0.5 mm apart. (See figures in Chapter 2 Astroecology.)

Table 2. Yields and statistical analysis of *Asparagus officinalis* tissue cultures grown on extracts of meteorites and simulants[a]

Material	Fresh weight (mg) Mean and std dev.	Sample vs. blank p values[b]	sample vs. DaG 476 p values[b]	sample vs. Murchison p values[b]	Number of samples
Blank water	0.23 *(0.03)*	--------	0.000	0.118	18
Allende set A	0.08 *(0.07)*	0.086	0.000	0.009	4
Murchison	0.32 *(0.04)*	0.064	0.050	--------	14
Theo's Flow Nakhla simulant	0.35 *(0.04)*	0.075	0.136	0.609	15
Hawaii lava Mars simulant	0.38 *(0.04)*	0.008	0.370	0.283	14
Basalt	0.43 *(0.06)*	0.008	0.936	0.149	6
DaG 476	0.44 *(0.04)*	0.000	--------	0.050	14
Allende set B	0.55 *(0.05)*	0.000	0.111	0.001	8
EETA 79001	0.58 *(0.20)*	0.000	0.020	0.000	11
MS nutrient medium	5.79 *(0.18)*	0.000	0.000	0.000	14
Reduced Light					
Allende set B	0.30 *(0.06)*	0.004[c]	1.000[d]	0.231[d]	6
Murchison	0.40 *(0.05)*	0.237[c]	0.231[d]		9
DaG 476	0.30 *(0.06)*	0.075[c]		0.231[d]	6

a. Yields (and standard errors) (mg) of *Asparagus officinalis* tissue cultures grown on extracts of meteorites or terrestrial analogues. Mann-Whitney non-parametric analysis p values. Mann-Whitney p values comparing the yields in the extract indicated by the row headings in full and reduced light.

b.   Mann-Whitney p values comparing the yields in the extracts indicated by the row and column headings, both grown in reduced light.

The statistical analysis may be applied to answer these questions: (1) Is any given treatment statistically different from blank water? (2) Are any two treatments significantly different from each other? (3) How does any given treatment compare with a full nutrient medium (Murashige and Skoog, 1962)?

As to the last question, all of the yields in Table 2 are much lower than in the full medium, and correspondingly, statistically distinct from the other media ($p = 0.000$). The lower yields in the extracts may reflect the absence, or unbalanced amounts, of nutrients.

In comparing the extracts with blank water, Table 2 shows increased yields with low p values, indicating that the extracts provided useable nutrients. The yields in the extracts of the two Martian meteorites were amongst the highest. Tables 2, columns 2 and 4 show that the DaG 476 extracts were most similar in yield and gave the highest p values when compared with the terrestrial basalt and lava ash, which are of similar mineralogical origin. The Murchison products showed the closest, possibly coincidental, similarity in yield and highest p value compared with the Theo's Flow cumulative igneous rock.

Tissue cultures of *Solanum tuberosum* (potato) were also observed. The Hawaii lava extracts produced strong green coloration, while the Murchison and DaG 476 extracts produced a brown coloration. The average fresh weights of these products were higher than the asparagus yields, in the order Hawaii Lava, 2.2; DaG 476, 1.8; Murchison, 1.7; Allende and Theo's Flow, 1.6 mg. However, none of the extract sets were statistically different from each other, with the minimum p value of 0.451 between any two sets. These observations are similar to previous results with variously supplemented Murchison extracts, where the potato cultures also gave higher weights but smaller statistical differences amongst the various media than asparagus (Mautner *et al.* 1997).

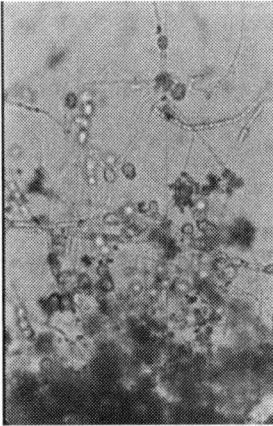

FIG. 2. Benthic growth of algae and fungi on powder of Dar al Gani 476 Mars meteorite. Yellow/green circles - *Chlorella;* white filament - fungal filaments; white spores - fungal spores. For scale, compare *Chorella* in Fig. 3.

Alternative bioassay agents may be plants grown from seeds. In the present work Arabidopsis *thaliana* was grown in the meteorite and rock extracts to plants of about 10 mm. In the first two weeks the plants grown in all the media were similar as they used the seed nutrients. After 4 - 6 weeks, the relative effects of the various extracts were qualitatively similar to those observed in the asparagus tissue cultures, except the plants in the extracts of Allende and Murchison that yellowed and degraded, suggesting toxic effects. Plants grown from seed merit further work as they may produce more uniform results than tissue cultures where it is hard to achieve uniform starting meristem tips.

Light intensity may affect the sensitivity of tissue cultures to the various media, and it will also be an important variable in planetary biology at varying heliocentric distances. The present cultures were developed at typical tissue culture growth chamber light fluxes of 80 μE m^{-2} s^{-1}, about 10% of the solar constant on Earth, as more intense light damages the cultures. This light flux is near the O_2 production/utilization compensation point. In relation to astrobiology, this light flux corresponds to the solar constant at about 3 au (asteroid belt). To test light flux effects, cultures of Murchison, Allende B and DaG 476 were grown in a partially shaded area of the growth chamber at a reduced flux of 12 μE m^{-2} s^{-1}, equivalent to the solar constant at about 9 au (Saturn).

Comparing the yields of each extract in full vs reduced light, the yields of the Allende B and DaG 476 extracts were smaller in reduced light than in full light, and the Man-Whitney analysis showed small p values in these comparisons, i.e., statistically meaningful effect of light reduction in these media (Table 2, columns 2 and 3, bottom rows). In Murchison extracts the light effects were statistically less significant.

It is also of interest if the light affects the differences between the yields in the various media. To examine this, the p values were calculated to compare pair wise the yields in Allende set B, Murchison and DaG 476 extracts, all in reduced light (Table 2, columns 4 and 5, bottom rows). These p values in reduced light were larger and showed less statistical distinctness than the yields in the same pairs of extracts in full light.

These observations suggest that light intensity is a significant variable in astrobiology. Nevertheless, the products observed in reduced light suggest that solar irradiation may support plant growth, although reduced, at the distances of the asteroid belt, and Jupiter and Saturn.

4. Algal Bioassays

Algae are the first colonizers in many the terrestrial environments and therefore also candidates for planetary terraforming. Algae are therefore relevant to planetary bioassays, and also convenient as their sizes allow microscopic analysis.

In the present experiments we used mesophilic algae isolated from soil in Canterbury, New Zealand. Preliminary experiments were also performed with cold-tolerant algae from sub-Antarctic islands. Mixed algal populations were used to allow a degree of natural selection, testing whether a dominant species will emerge or a diverse stable ecosystem is established in the microcosms.

The algae were identified microscopically to genus level as *Leptoyngbya sp.* (filamentous blue-green), *Klebsormidium sp.* and *Stichococcus sp.* (filamentous green), *Chlorella sp.* (green unicellular), *Chlorosarcinopsis sp.* (green unicellular in aggregates), and *Navicula sp.* (diatom). The inoculant cultures also contained flagellate bacteria of about 1 micron diameter.

In the cultures grown on cavity slides as described above, the meteorite extracts and solids served as minaturized planetary microcosms simulating small ponds. The growth of the algae was examined microscopically at weekly intervals. Most of the growth of the filamentous blue-green algae and *Chlorella* occurred as benthic growth.

Green cells and filaments were observed to survive for at least four weeks in most extracts. The best growth, illustrated in Fig. 2, was observed with the DaG 476 extract and solids. In contrast, in the Allende A, materials growth ceased after 6 – 10 days and only shells of algae were observed later, consistent with the inhibitory effects of this material noted above. Fig. 2 also shows a dense growth of fungal filaments and spores. Since the DaG 476 material is not known to contain organics, the fungal growth appears to be supported by algal detritus, and enhanced by the high extractable phosphate content of this meteorite.

The benthic growth did not allow the microscopic counting of cell populations because of the interference of the solids. For quantitative measurements cultures were grown in extracts in vials, without the solids present, and counted in a haemocytometer chamber. Unicellular *Chlorella* as single cells and some *Chlorosarcinopsis* in aggregates of 2 – 20 cells were dominant in all the extract cultures. A few filaments, 8 – 40 micrometers in length and containing 2 – 10 cells were observed, as well as occasional short *Klebsormidium* filaments and *Stichococcus* fragments, with cell populations smaller by an order of magnitude than the unicellular algae. Fig. 3 shows examples of the algal populations obtained after 32 days. The unicellular chlorophytes dominated in all the cultures, but the diversity of organisms varied in the different extracts. Fungal spores were also observed especially in the Murchison extracts, where they may be supported by the organic content of the meteorite. This is similar to the efficient growth of heterotrophic bacteria and fungi observed in other cultures of Murchison extracts (Mautner 1997a). Although the starting inoculants contained diatoms, no diatom populations were observed after 1 – 4 weeks.

Growth curves for cultures without added nitrate are shown in Fig. 4. The relative cell populations after 34 days followed a similar order as in the plant tissue cultures. Here also the yields of DaG 476 exceeded Hawaii lava and Murchison, while the Allende A extracts gave the lowest populations.

The population levels were higher in the cultures containing added nitrate, but the relative populations in the various extracts followed a similar order to that without nitrate. The populations grew for about 8 days and remained approximately constant for a further 8 - 20 days. After 8 days the product populations were DaG 476, $1.1x10^6$; Murchison, $0.30x10^6$; Hawaii lava, $0.27x10^6$; Theo's Flow, $0.23x10^6$; Allende A, $0.12x10^6$ cells mL^{-1} . These relative yields were consistent with the trends in the extracts without added nitrate and in the plant tissue cultures with added nitrate. The fact that added nitrate had little effect on the relative yields suggests that phosphate rather than nitrate was the limiting nutrient.

Cold tolerance is required for potential growth on the asteroids and outer planets. For this reason, preliminary experiments were performed with cold-tolerant *Chlorella sp.*, *Stichoccus sp.* and filamentous blue-green *Oscillatoriaccie sp.* collected from sub-Antarctic islands. The cultures were grown in meteorite extracts at 12 °C inoculated with 10^4 CFU mL^{-1} of each alga. After 60 days, the combined populations of *Chlorella* and *Stichoccus* were: Nutrient solution, 25.9; Basalt, 6.1; Allende B, 2.5; Hawaii lava, 1.5; Lunar Simulant, 1.3; DaG 476, 1.1; Theo's Flow, 1.0; Murchison, 0.9; Templeton soil, 0.6; blank water control, 0.2 x10^6 CFU mL^{-1}. In these cultures also, the filamentous blue-green algae, here *Oscillatoriaccie* did not grow well and disappeared after 20 days.

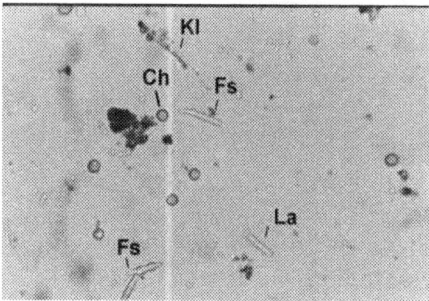

Murchison Meteorite

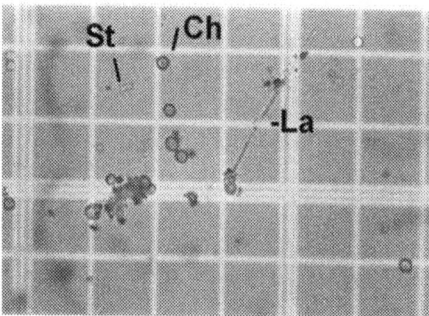

DaG 476 Martian Meteorite

FIG. 3. Algal populations observed in meteorite and soil extracts supplemented by 0.5 mM NH$_4$NO$_3$ after 32 days of growth. (Top) Murchison. (Bottom) DaG476. Ch - *Chlorella*; Kl - *Klebsormidium*; La - *Leptolyngya*; St - *Stichococcus*; Fs - Fungal spores. The spacing between the counting chamber lines is 0.2mm.

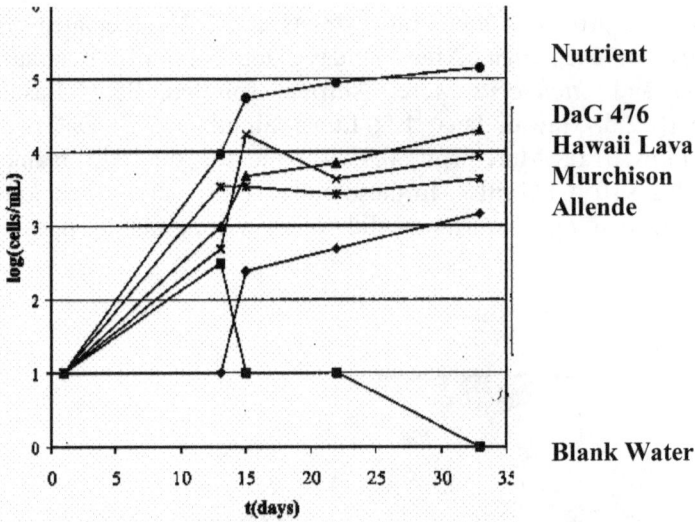

FIG. 4 Algal population growth curves in meteorite and simulant extracts and BG-11 nutrient medium. Note the order of yields: BG-11 nutrient medium > DaG 476 Martian meteorite > Hawaii lave Mars simulant > Murchison > Allende A > deionized water.

In addition to the algal growth, all of the cultures contained motile microorganisms of about 1 μm diameter. The bacteria attained observable populations after 2 - 4 weeks, and at later times become numerically the largest population, at the ratio of 2 – 4 microorganisms per algal cell. The populations were highest in the extracts Templeton Soil, Murchison and Hawaii lava, in correlation with the organic contents of the solids. When cultured on nutrient or potato agar in dark, the microorganisms yielded bright orange colonies, but not when the agar was supplemented with chloretetracyclin (auremycin) antibacterial agent. The microorganisms were motile heterotrophic bacteria supported by organics in the extracts and by the algal metabolites and decay products.

Discussion

1. Design Considerations For Planetary Microcosms

The design of planetary microcosms depends on the objectives of the simulation, such as the testing of planetary materials in relation to
244

early life or as future ecosystems. Early planetary conditions may have been reducing or dominated by carbon dioxide. On the other hand, terraforming will aim to create habitable oxygen-rich environments. Our present tests address the latter environment, but anaerobic tests are also under way.

To assess a complex planetary ecosystem, it is necessary to examine the mutual effects of atmospheric, aqueous and geological processes. Such studies may be limited, however, by the available amounts of meteorites. For example, the analysis of extracted nutrient cations commonly applies atomic absorption spectroscopy or ICP-MS, and for anions, ion exchange chromatography. These techniques require about 1 mL solutions containing 10^{-6} g of solute, usually from solids that contain them in soluble forms mostly in concentrations of $10^{-3} - 1$ g kg^{-1} (see Table 1), i.e., requiring $0.001 - 1$ g mineral samples. Phosphate can be quantified reliably using colorimetry in 1 mL samples at concentrations of 10^{-5} g L^{-1} , and is usually contained at extractable levels of 10^{-4} to 10^{-2} g kg^{-1} (see Table 1), requiring 0.001 to 0.1 g mineral samples.

For biological tests, the microcosm must contain enough nutrients to support the population. In plant bioassays, the sample plants from tissue cultures yield typically fresh weights of 10^{-3} g containing 1 g kg^{-1} or a net 10^{-6} g of macronutrients such as Ca or K. This requires 0.001 - 1 g of mineral sample per plant, and usually $4 - 10$ plants per experiment to allow statistical comparisons (Table 3a and 3b).

For microbial or algal microcosms, population levels of 10^7 cells/mL in 1 mL cultures need to be supported. With an average algal mass of 10^{-10} g/cell again a macronutrient content 10^{-6} g extracted from $0.001 - 1$ g of solid, may be required. For tests using bacteria, populations of 10^7 cells/mL of microorganisms of mass of 10^{-13} to 10^{-12} g/cell, the requirements are smaller, $10^{-9} - 10^{-8}$ g of nutrient extracted from $10^{-6} - 10^{-2}$ g of solid.

The material requirements for chemical analysis and plant and algal microcosm cultures are therefore usually $0.001 - 1$ g per test. A comprehensive ecological study of a microcosm may require ten or more chemical and biological analyses and 0.01 - 10 g of material. These considerations illustrate that microcosm simulations are possible with small amounts of meteorite materials, but the experiments must be designed judiciously.

Table 3a. Relative algal and plant culture yields[a] of meteorites and analogous rocks

	Mesophile algae[a,b]	Antarctic algae[a,c]	Aspara-gus[a]	Potato[a]	Mean relative yield[d]
Allende B		8.4	5.5	7.6	7.2
Murchison	3.1	2.4	1.6	8.1	3.8
DaG 476	14.1	3.1	3.6	9.0	7.4
EETA 79001	14.4		6.0		10.1
Basalt	14.2	22.4	3.6		13.3
Hawaii lava	4.7	4.7	2.6	11.1	5.8
Lunar Simulant	0.0	3.7		6.0	3.2
Theo's Flow	0.7	2.7	2.1	7.7	3.3
Templeton Soil	18.6	1.2			9.9
cv[g]	0.84	0.62	0.48	0.21	0.49

Footnotes to Table 3a

a. Yields of algal populations and tissue culture products, normalized to yields in optimized media (see equation (2) and following text).
b. Normalized algal populations after 34 days.
c. Normalized algal populations after 60 days.
d. Mean over columns j = 2 - 5 of the normalized yields x_{ij} for material i.

Table 3b. Limiting nutrients and a bioassay ranking of meteorites and analogous rocks.

	Rating by algal yield[e]	Rating by average yield[e]	Rating by NO_3-N[e]	Rating by PO_4-P[e]	Mean Z score and fertility rating[f]
Allende B	+	++	+	+	-0.22 M
Murch-ison	+	+	++	+	-0.57 M
DaG 476	++	++	+++	++	0.58 H
EETA 79001	+++	++	+++	+++	1.32 VH
Basalt	+++	+++	+	++	0.65 H
Hawaii lava	+	+	+	O	-0.56 M
Lunar Simulant	O	O	O	++	-0.36 M
Theo's Flow	O	O	+	O	-1.01 L
Templeton Soil	++	++		+	0.21 H

Footnotes to Table 3b.

e. Rated using the standard normal variate (Z score = $(x_{i,j} - \mu_j)/\sigma_j$) (see equation (3) and following text).
f. Average of the $Z_{i,j}$ values for each material, and rating based on this value (see text).
g. Coefficient of variation cv = std/mean of the values in each column.

Assessing Relative Fertilities
(Calculating the 'Z' Score)

The Z score is obtained by calculating the average performance of all the materials in a given test, and then calculating the deviation of each given material from the average. The deviation for a particular material is more meaningful the less scattered are the other results. Therefore the difference is divided by a measure of the scatter, the standard deviation (std). The Z score for property x_i of material i is $Z_i = (x_i - x_{average})/std$.

For example, the extract of the Murchison meteorite yielded a deviation of 3.1% and the DaG 476 Martian meteorite yielded 14.1% compared with algal populations in an optimized growth medium. The average yield was 8.7% of the optimized yield, with a std of 6.9%. Accordingly, the Murchison meteorite received a Z score of –0.82, and DaG 476 received a Z score of 0.78 in the algal test. The other biological and nutrient tests were evaluated similarly.

For a more qualitative rating, the Z score of each material in each test was converted in this manner: a Z score greater than 1 was assigned a +++ fertility rating; between 0 and 1, ++; between 0 and -1 rated +; and less than –1, rated as O. We rated all of the materials based on: the growth of mesophilic algae (that prefer moderate temperatures); cold-adapted Antarctic algae; yields of potato and asparagus tissue cultures; and the extractable nitrate and phosphate nutrient contents of each material. Finally, the results of all the tests were combined in the average Z score of each material in all the tests, and each material was assigned a rating of very high, high, medium or low fertility on this basis. Table 2.1 shows the results of these assignments.

2. Simulated Solutions In Asteroid, Cometary and Meteorite Interiors

Extracts obtained under planetary conditions can simulate interior solutions in meteorites and their parent bodies during aqueous alteration or after infall to aqueous early planets. In fact, the high soluble salt contents in Murchison may have been deposited during the alteration process, after the water evaporated or was absorbed by serpentine formation (Tomeoka and Buseck 1985, Bodnar and Zolensky 2000, Cohen and Coker 2000). Similarly, high levels of soluble salts in the Nakhla Martian meteorite were considered to be evaporates from a salty ocean, (Sawyer *et al.* 2000) and this may also apply to DaG 476.

To simulate these interior solutions, the solids should be extracted at the high solid/solution ratios that apply in the internal pores. In Murchison with a porosity of 23% this requires a solid/solution ratio of about 10 kg L^{-1}. Extractions in the range of 1 - 10 kg L^{-1} showed that the amounts extracted per gram solid remain constant in this range of solid/solution ratios, at the values listed Table 1. For extractions at $r_{solid/solution}$ = 10 kg L^{-1} the resulting solution concentrations in mol L^{-1} can be obtained by multiplying the values in Table 1 by 10/MW (molecular weight of the solute). The result are 0.4, 0.7, 0.8 and 0.03 mol L^{-1} for the cations Ca^{2+}, Mg^{2+}, Na^+ and K^+ and 0.05, 0.009, 1.8 and 0.002 mol L^{-1} for Cl^-, NO_3^-, SO_4^{2-} and PO_4^{3-}, respectively. The total measured ion concentration is 3.8 mol L^{-1} with an ionic charge of 6.7 equivalents L^{-1}. Similarly, from the amount of soluble organics one can estimate a concentration of organics (using average MW = 100) as 0.1 mol L^{-1} (Mautner *et al.*, 1995).

From a biological point of view, trace elements are significant as micronutrients. Extractions of Murchison in water at $r_{solid/solution}$ = 0.021 kg L^{-1} yielded only Mn at 0.005 and Ni at 0.1 g kg^{-1}, while more aggressive extraction in 1 M ammonium acetate yielded B, 0.0005; Fe, 0.009; Mn, 0.011; Al, 0.00003; Cd, 0.0038; Cr, 0.0002; Cu, 0.0005; Ni, 0.1 g kg^{-1}. Converted to solution concentrations at the solid/solution ratios of 10 kg L^{-1} applicable in asteroid pores, all the concentrations are in the range found in of soil solutions and can be sufficient as micronutrients. However, the high concentration of Ni may be toxic.

The calculated concentrations show that the pores of carbonaceous chondrite parent asteroids and CM2 meteorites may contain highly concentrated electrolytes and organics under aqueous conditions. Can microorganisms grow in these solutions? As a test, samples of Murchison that were wetted at solid/solution ratios of 10 kg L^{-1} were inoculated with microbial isolates from a wetland. The cultures produced microbial communities of several algal species, bacteria assigned tentatively as *Clavibacter michganenese, Microbacterium imperiale, Eurobacterium saperdae, Pseudomonas putida* and *Corynebacterium sp.,* and a filamentuos fungus. These diverse microbial populations grew and survived in these concentrated meteorite/asteroid interior solutions for over six months.

3. An Assessment of The Biological Test Agents

It is of interest to correlate the responses of the various test organisms to each other, and to correlate these biological responses with the nutrient contents of the various media.

The algal populations produce yields in units of CFU/mL and the plant cultures in units of plant mass in mg. To compare these responses on a common basis, the yield of each culture in medium X was normalized using equation (2) to the yield in the respective optimized medium.

Yield X (%) = 100 [yield (X)- yield (blank water)] /
[yield (in optimized medium)] (2)

The algal populations were normalized to those obtained in BG-11 medium (Rippka 1979) and the plant fresh weights to those obtained in an MS medium (Murashige and Skoog 1962). The yields in water and in optimized media used for the normalization were: mesophile algae, 0 and 1.4E5 CFU/ml; Antarctic algae, 2.9E5 and 2.6E7 CFU/ml; asparagus, 0.23 and 5.79 mg; potato, 0.2 and 18.0 mg. The normalized yields as shown in Table 3 are low but they show differences amongst the various extracts. The overall biological response to each extract is reflected in Table 3, column 6 by the average of the normalized yields.

It is of interest whether the relative responses to the various media were correlated amongst the various test organisms. Pearson correlation analysis of the yields showed the highest correlation coefficient between the yields of mesophile algae and asparagus cultures (cor. coeff. = 0.716) and the smallest correlation coefficient between the yields of the Antarctic algae and potato cultures (cor. coeff. = 0.005).

A discriminating biological assay agent should give distinct responses to different media. In this respect, a large scatter of the responses of a given organism to various media indicates the ability of that organism to discriminate amongst the media. A measure of this scatter is the coefficient of variation (cv = std/mean) of the set of responses of each assay agent j (columns 2-5, Table 3) to the set of media in rows i, with the resulting cv values for each test agent j shown in the last row of Table 3. By this measure the organisms applied are discriminating in the order of cv values: Mesophile algae (0.84) > Antarctic algae (0.62) > asparagus (0.48) > potato (0.21).

4. Relation Between Nutrients And Biological Responses

The responses of the various test organisms should correlate with the nutrients in the mineral extracts. To test this a Pearson correlation analysis was carried out between the normalized yields of each algal and plant test organism in Table 3 and the nutrient concentrations derived from the data in Table 1. The biological yields showed negative correlations with most of the electrolytes in the extracts, i.e., Ca, Mg, Na, K, Cl and SO_4-S, with correlation coefficients mostly between -0.2 and -0.5, but positive correlations with the NO_3-N and PO_4-P concentrations. The largest positive correlation coefficients were obtained with the yields of mesophilic algae, that showed correlation coefficients of +0.28 with NO_3-N and +0.32 with PO_4-P. The asparagus yields also gave positive correlation coefficients of +0.31 with NO_3-N and +0.74 with PO_4-P. The results indicate that PO_4-P and to a lesser degree NO_3-N are the limiting nutrients.

The overall biological response is measured by the combined average normalized yields of the algae and plants in Table 3, column 6. An analysis of these yields vs. the concentrations of nutrients gave the correlation coefficients Ca, -0.38; Mg, -0.36; Na, -0.34; K, -0.36; Cl, -0.37; NO_3-N, +0.08; SO_4-S, -0.36; PO_4-P, +0.35. These results again suggest that nitrate and especially phosphate are the liming nutrients in these materials.

An alternative way to identify the limiting nutrients is to compare the relative amounts of the nutrients in each medium with the relative amounts of these nutrients in biomass. For example, the concentration of nutrient X_i in extract i may be normalized to the limiting nutrient (NO_3-N) in the same extract. The normalized concentrations ($[X]/[NO_3$-N$])_i$ in solution can be compared with the normalized concentration of the same nutrient ($[X]/[NO_3$-N$])_{biomass}$ in algal biomass (Bowen, 1966). The ratios of the normalized concentrations $[X/(NO_3$-N$)]_i/(X/((NO_3$-N$))_{biomass}$ show the excess of the element in extract i over the amount needed to construct algal biomass when all the NO_3-N in the extract is converted to biomass. The comparison shows, for example, that if all the NO_3-N extracted from EETA 79001 is converted to plant biomass, Ca in this extract is in excess by a factor of 23, Mg, 59; Na, 145; K, 2.6; Cl, 42; SO_4-S, 34; PO_4-P, 252, over the amount required to complement NO_3-N. Similarly, if the available NO_3-N in Murchison is converted to algal biomass, Ca is in excess by a factor of 493; Mg, 1,428; Na, 80; K, 12; Cl, 177; SO_4-S, 1188; and PO_4-P, 3.5 to form algal biomass. This analysis further suggests that NO_3-N is the limiting nutrient in the Martian meteorites while in Murchison both NO_3-N and PO_4-P are the limiting nutrients. Of course, this analysis does not include C, which is available on Mars from the atmosphere and in Murchison from the 2% organic fraction.

The high fertility of the Martian meteorites in the final ranking in Table 3 may be due primarily to the relative high extractable nitrate and phosphate. Conversely, Murchison ranks only medium despite its high content of extractable electrolytes, probably because of the relatively low levels of nitrate and phosphate.

5. A Bioassay Of Martian And Meteorite Materials And Analogues

The results of the biological tests and nutrient analysis can be combined for an overall rating of the biological potentials of the tested materials. The rating is based on the average algal yields, plants yields, combined average biological yield, and concentrations of the limiting nutrients NO_3-N and PO_4-P. These different criteria must be accounted for in a form that allows comparisons among tests measured in different units. In this work, a standard variate Z score was applied, that relates the result of each test on material i to the average and std of the same test over all the materials, using equation (3).

$$Z_{i,j} = (x_{i,j} - \mu_j)/\sigma_j \qquad (3)$$

Here x_{ij} is the value of result j (biological yield, columns 2 - 6 in Table 3 or nutrient concentration, columns 8 and 10 in Table 1) for a given material i (rows 1 - 11 in Table 3), and μ_j and σ_j are the mean and std of results of test j averaged over all the materials. The resulting Z scores were then used to rate material i according to result j, as follows: $Z_{ij} > 1$, rated +++; $0 < Z_{ij} < 1$ rated ++; $-1 < Z_{i,j} < 0$ rated +; $Z_{ij} < -1$ rated O. For example, the average of the mean normalized biological yields in Table 3, column 6 is $\mu_j = 7.11\%$, with an std of $\sigma_j = \pm 3.49\%$. Using equation (3), these values combined with the normalized biological yield of 10.1% of EETA 79001 gave a Z score of 0.86, which was rated as ++ (Table 1, column 8).

The criteria rated in this manner in Table 3 are the mean algal yield (average Z score for mesophile and Antarctic algae, Table3, columns

2 and 3), the overall mean biological (algal and plant) yield (Table 3, column 6), and the extractable NO_3-N and PO_4-P concentrations from Table 1. Finally, an overall rating is assigned to each material i based on the mean of the Z scores for material i over the criteria j applied, as shown in Table 3, column 11. The ratings VH, H, M or L were assigned to the fertility of the material on the same basis as the +++ etc. values assigned in the preceding paragraph. For example, the average of the Z scores for EETA79001 in the various tests was 1.32, which was rated as VH.

The various fertility criteria obtained by this procedure in Table 3 show internal consistency. The basaltic Martian meteorites and the terrestrial basalt sample rate ++ or +++ in most individual tests and VH or H in the overall evaluation. The two carbonaceous chondrites Allende and Murchison rate + or ++ in the tests and M overall. The two lava ash NASA simulants, JSC Mars-1 the JSC-1 Lunar Simulant, and the Theo's Flow Nakhla analogue rate O or + in the tests and M in the overall rating. The tests therefore assign comparable ratings to related materials. The Martian materials rate higher than their proposed analogues, suggesting that biological simulants should be selected by different criteria than mineralogical or physical simulants. The result that the fertility tests group similar materials together suggests that the tests are internally consistent.

Conclusions

1. Microcosm Bioassays

The present work tests the application of planetary microcosms to assay the biological fertilities of extraterrestrial materials, using small samples of 10 - 100 mg material suitable for meteorite studies. The procedure in this work consists of:

(1) Aqueous extraction, and analysis, of nutrient electrolytes at solid/solution ratios of 0.01 - 10 kg L^{-1}, using 0.1 kg L^{-1} in most cases.

(2) Biological tests using mixed algal populations that allow a degree of natural selection, and tests using plant tissue cultures.

(3) Correlation analysis amongst the biological responses to various meteorites, to test the consistency of the bioassays.

(4) Using the bioassay to identify the most discriminating and sensitive test organisms.

(5) Correlation analysis between the overall biological responses and extracted nutrients to identify the critical limiting nutrients.

(6) Based on the extractable nutrients and biological responses, assigning a fertility ranking to various materials.

The various tests applied were generally consistent. Mesophile green algae and asparagus tissue cultures were found to be sensitive and discriminating bioassay agents. Testing the correlation between nutrients and biological yields and suggests that NO_3-N and PO_4-P are the limiting nutrients in the meteorites. The fertility ranking groups materials in the order Martian basalts > terrestrial basalts, agricultural soil > carbonaceous chondrites, lava ash > cumulate igneous rock. The tests rate materials of similar mineralogy together in terms of fertility. The relation between the relative biological yields and the extractable nutrients is also reasonable. These observations suggest that the bioassays are internally consistent.

2. Implications For Astrobiology And Astroecology

Astrobiology concerns the past evolution and potential future of life in the universe (Morrison 2001). The microcosm results are relevant to these areas, although the present small microcosm models can provide only a few indications of the potential applications.

Based on the present results, carbonaceous asteroids during aqueous alteration and meteorites after infall to aqueous planets, can form concentrated internal solutions of nutrient electrolytes (3.8 mol L^{-1} electrolytes) and soluble organics (0.1 mol L^{-1}). These trapped solutions can allow the multi-step synthesis of complex organics, contributing to biogenesis. The observed microbial and algal growth suggest that these solutions can also support early microorganisms. These observations may support tentative observations of Dinoflagellate and Chysomonad algal fossils in carbonaceous chondrite meteorites (Claus and Nagy 1961, Urey 1962). If nutrient solutions and microorganisms are indeed present in early aqueous carbonaceous asteroids, frequent collisions amongst these objects may facilitate the growth and dispersion of the microorganisms in solar nebulae.

The results show that light intensity is a significant variable in astrobiology at various heliocentric distances. The observed yields in reduced light suggest that solar irradiation can support plant growth, although at reduced yields, even at large heliocentric distances corresponding to the asteroid belt, and Jupiter and Saturn and their moons.

The results on nutrients allow estimating the biomass sustainable by planetary materials, for example, carbonaceous chondrite asteroids that were proposed for as space resources (O'Neill 1974, O'Leary 1977, Lewis 1993, Sagan 1994, Dyson 2000). Based on Table 1, the total 10^{22} kg of carbonaceous chondrite material in the asteroid belt (Lewis 1997) contains 8×10^{16} kg extractable NO_3-N and 5×10^{16} kg extractable PO_4-P. If distributed as synthetic soils, these materials can sustain up to 10^{18} kg microbial or plant biomass, compared with the estimated terrestrial biomass of 10^{15} kg (Bowen, 1966). Similar carbonaceous chondrite materials in Phobos and Deimos may also be useful soil resources. The high extractable NO_3-N and PO_4-P contents of the Martian meteorites suggest that igneous Martian rocks can be processed into useful soils for terraforming. The large sustainable biomass by these various materials suggests that planetary resources can accommodate substantial biological activity.

In summary, planetary microcosms, based on meteorite materials can be useful in experimental astroecology. Larger samples from sample return missions, or in situ studies in planetary locations will allow simulations with larger sets of bioassay agents. Such comprehensive microcosm studies can monitor simultaneously the evolution of nutrients and biota, to reveal links amongst these ecological variables. The present studies using the limited available materials suggest that microcosm studies can provide consistent bioassays. The resulting fertility ranking of planetary materials can help in targeting astrobiology searches for past life, and in identifying planetary bioresources.

Acknowledgements. I thank Prof. Anthony J. Conner for plant tissue culture samples, Dr. Paul Broady for algal cultures and taxonomical identification, Mrs. Helene D. Mautner for assistance with the plant and algal studies, The Smithsonian Institute for samples of Allende and Murchison, Dr. Carleton Allen and the NASA Johnson Space Center for simulated Mars and Lunar soils and samples of the EETA 79001 meteorite, Mr. Gavin Robinson for the ICP-MS analysis, and Prof. K. C. Cameron and Drs. Robert Sherlock, Robert Leonard and Eric Forbes for helpful discussions, and Dr. Chris Frampton for help with the statistical analysis. The comments of an anonymous reviewer pointed out the significance of light intensity effects. This work was funded by grant LIU 901 from the Marsden Foundation, administered by the Royal Society of New Zealand.

References

1. Allen, C. C., R. V. Morris, K. M. Jager, D. C. Golden, D. J. Lindstrom, M. M. Lindstrom and J. P. Lockwood 1998. Martian regolith simulant JSC MARS-1. *Lunar and Planetary Science* **XXIX.**

2. Arrhenius, S. 1908. *Vernaldas Ultveckling*. Stockholm.

3. Barber, D. J. 1981. Matrix phyllosilicates and associated minerals in CM2 carbonaceous chondrites. *Geocim. Cosmochim. Acta* **45,** 945-970.

4. Beare, M. H., K. C. Cameron, P. H. Williams and C. Doscher 1997. Soil quality monitoring for sustainable agriculture. *Proc. 50th New Zealand Plant Protection Conf.* 520-528.

5. Bodnar, R. and M. Zolensky 2000. Liquid-water fluid inclusions in chondritic meteorites: Implications for near-surface P-T conditions on parent asteroids. *Journal of Conference Abstracts* **5,** 223.

6. Bowen, H. J. M. 1966. *Trace Elements in Biochemistry*. Academic Press, New York, pp. 68.

7. Brierly, and Jones 1998. Carbonaceous meteorites. In *Planetary Materials* (J. J. Papike, ed.) Reviews in Mineralogy v. 36, Mineralogical Society of America.

8. Bunch, T. E. and S. Chang 1980. Carbonaceous chondrites. II. Carbonaceous chondrite phyllosillicates and light element geochemistry as indicators of parent body processes and surface conditions. *Geochim. Cosmochim. Acta* **44,** 1543-1577.

9. Chyba, C. F., and C. Sagan 1992. Endogenous production, exogenous delivery and impact-shock synthesis of organic molecules: An inventory for the origins of life. *Nature* **335,** 125-132.

10. Chyba, C. F. and G. D. McDonald 1995. The origin of life in the Solar System: Current Issues. *Annual. Rev. Earth Planet. Sci.* **23,** 215-249.

11. Claus, G., and B. Nagy 1961. A microbiological examination of some carbonaceous chondrites. *Nature* **192,** 594-598.

12. Cohen, B. A. and R. F. Coker 2000. Modelling of liquid water on CM meteorite parent bodies and implications for amino acid racemization. *Icarus* **145,** 369-381.

13. Crick, F. H. and L. E.Orgel 1973. Directed panspermia. *Icarus* **19,** 341-348.

14. Dyson, F. 2000. *Imagined Worlds - Jerusalem - Harvard Lectures*. Harvard University Press, Cambridge.

15. Friedman, R. C. 1998. Petrologic clues to lava flow emplacement and post-emplacement process. Ph. D. Thesis, university of Hawaii, Department of Geology and Geophysics.

16. Frossard, E., and S. Sinaj 1997. The isotope exchange kinetic technique: A method to describe the availability of inorganic nutrients. Applications to K, P, S, and Zn. *Isotopes in Environ. Health Stud.* **33**, 61-77.

17. Fuchs, L. H., E. Olsen and K. J. Jensen 1973. Mineralogy, mineral-chemistry, and composition of the Murchison (C2) meteorite. *Smithsonian Contributions to the Earth Sciences, number 10.* pp. 1-39. Smithsonian Institution Press, Washington, D. C.

18. Greenberg, J. M., and Li, A. 1998. Comets as a source of Life's origins. In *Exobiology: Matter, Energy, and Information in the Origin and Evolution of Life in the Universe.* (J. Chela-Flores and F. Raulin, eds.) Kluwer Academic Publishers. pp. 275-285.

19. Jarosewich, E. 1971. Chemical analysis of the Murchison meteorite. *Meteoritics* **1**, 49-51.

20. Kasting, J. F. and M. A. Mischna 2000. The influence of carbon dioxide clouds on early martian climate. *First Astrobiology Science Conference,* NASA Ames Research Center, Moffett Field, California. April 2000. p. 24.

21. Komacki, A. S., and J. A. Wood 1984. The mineral chemistry and origin of inclusion matrix and meteorite matrix in the Allende CV3 chondrite. *Geochim. Cosmochim. Acta* **48**, 1663-1676.

22. Leslie, A. 1879. *The Arctic Voyages of Adolf Eric Nordenskiold.* McMillan and Co., London. P. 65. Refers to observations of algae by the botanist Dr. Berggren on cryoconite, later identified as micrometeorite dust (Maurette *et al.* 1986*).*

23. Lewis, S. J. 1993. *Resources of Near-Earth Space.* University of Arizona Press, Tucson, Arizona.

24. Lewis, S. J. 1997. *Physics and Chemistry of the Solar System.* Academic Press, New York.

25. Lovering, J. F., R. W. Le Mai and B. W. Chappell 1971. Murchison C2 carbonaceous chondrite and its inorganic composition. *Nature Physical Science* **230**, 14-20.

26. McKay, D. S., J. L. Carter, W. W. Boles, C. C. Allen and J. H. Alton 1993. JSC-1: A new lunar regolith simulant. *Lunar and Planetary Science* **XXIV**, 963-964.

27. McKay, D. S., K. L. Gibson, H. Thomas-Kerpta, C. S. Vali, S. J. Romaneck, X. D. F. Clemett, C. R. Chillier, C. R. Maechling, and R. N. Zare. 1996. Search for past life on Mars: Possible relic biogenic activity in the martian meteorite ALH84001. *Science* **273**, 924-930.

28. McLaren, R. G. and K. C. Cameron 1996. *Soil Science.* Oxford University Press, Auckland.

29. McSween, H. Y. Jr. and E. Jarosewitz 1983. Petrogenensis of the Elephant Moraine A79001 meteorite: Multiple magma pulses on the shergottite parent body. *Geochim. Cosmochim Acta* **47**, 1501-1513.

30. McSween, H. Y., Jr. and A. H. Treiman 1998. Martian Meteorites. In *Planetary Materials* (J. J. Papike, ed.) Reviews in Mineralogy v. 36, Mineralogical Society of America

31. pp. 6-1 – 6-54.

32. Maurette, M., C. Hammer, D. E. Brownlee, N. Reeh and H. H. Thomsen 1986. Placers of cosmic dust in the blue ice lakes of Greenland. *Science* **233**, 869-872.

33. Mautner, M. N. 1997 a. Biological potentials of extraterrestrial materials. 1. Nutrients in carbonaceous meteorites, and effects on biological growth. *Planetary and Space Science* **45**, 653-664.

34. Mautner, M. N. 1997 b. Directed panspermia. 3. Strategies and motivation for seeding star-forming clouds. *J. British Interplanetary Soc.* **50**, 93-102.

35. Mautner, M. N. 1999. Formation, chemistry and fertility of extraterrestrial soils: Cohesion, water adsorption and surface area of carbonaceous chondrites. Prebiotic and space resource applications. *Icarus* **137**, 178 – 195.

36. Mautner, M. N. and S. Sinaj 2002. Water-extractable and exchangeable phosphate in Martian and carbonaceous meteorites and planetary soil analogues. Submitted for publication.

37. Mautner, M. N., R. L. Leonard and D. W. Deamer 1995. Meteorite organics in planetary environments: Hydrothermal release, surface activity, and microbial utilization. *Planetary and Space Science* **43**, 139-147.

38. Mautner, M. N., A. J. Conner, K. Killham and D. W. Deamer 1997. Biological potential of extraterrestrial materials. 2. Microbial and plant responses to nutrients in the Murchison carbonaceous meteorite. *Icarus* **129**, 245-253.

39. Mileikowsky, C., F. A. Cucinotta, J. W. Wilson, B. Gladman, G. Horneck, L. Lindegren, J. Melosh, H. Rickman, M. Valtonen and J. Q. Zheng 2000. Natural transfer of viable microbes in space. 1. From Mars to Earth and Earth to Mars. *Icarus* **145**, 391 - 428.

40. Ming, D. W. and D. L. Henninger, (Eds.) 1989. *Lunar Base Agriculture: Soils for Plant Growth.* Amer. Soc. for Agriculture, Madison.

41. Morrison, D. 2001. The NASA astrobiology program. *Astrobiology* **1**, 3-13.

42. Murashige, T. and F. Skoog 1962. A revised medium for rapid growth and bioessays with tobacco tissue cultures. *Physiol. Plant* **15**, 473-497.

43. O'Leary, B. T. 1977. Mining the Apollo and Amor asteroids. *Science* **197**, 363-364.

44. O'Neill, G. K. 1974. The colonization of space. *Physics Today* **27**, 32-38.

45. O'Neill, G. K. 1977. *The high frontier.* William Morrow, New York.

46. Rippka, R. 1979. *J. Gen. Microbiol.* **111**, 1-61.

47. Sagan, C. 1994. *Pale Bule Dor: A Vision of Human Future in Space.* Random House, New York.

48. Sawyer, D. J., M. D. McGehee, J. Canepa and C. B. Moore. 2000. Water soluble ions in the Nakhla martian meteorite. *Meteoritics and Space Science* **35**, 743 – 747.

49. Stoker, C. R., P. J. Boston, R. L. Mancinelli, W. Segal, B. N. Khare and, C. Sagan. 1990. Microbial metabolism of tholin. *Icarus* **85**, 241-248.

50. Steele, A., K. Thomas-Kerpta, F. W. Westall, R. Avci, E. K. Gibson, C. Griffin, C. Whitby, D. S. McKay, and J. K. W. Toporski. 2000. The microbiological contamination of meteorites: A null hypothesis. *First Astrobiology Science Conference.* NASA Ames Research Center, Moffett Field, California, April 2000. p. 23.

51. Tomeoka, K. and P. R. Buseck 1985. Indicators of aqueous alteration in CM carbonaceous chondrites: Microtextures of a layered mineral containing Fe, S, O and Ni. *Geochim. Cosmochim. Acta* **49**, 2149-2163.

52. Urey, H. C. 1962. Lifelike forms in meteorites. *Science* **137**, 625-628.

53. van Veldhoven, P. P. and G. P. Mannerts 1987. Inorganic and organic phosphate measurements in the nanomolar range. *Anal. Biochem.* **161**, 45-48.

54. Zipfel, J., P. Scherer, B. Spettel, G. Dreibus and L. Schultz 2000. Petrology and chemistry of the new shergottite Dar al Gani 476. *Meteoritics and Planetary Science* **35**, 124-128.

Further Papers on Astroecology:

Mautner, M. N.; Leonard, R. L.; Deamer, D. W. "Meteorite organics in planetary environments: Hydrothermal release, surface activity, and microbial utilization". Planetary and Space Science, 1995, 43, 139.

Mautner, M. N. "Biological potential of extraterrestrial materials. 1. Nutrients in carbonaceous meteorites, and effects on biological growth." Planetary and Space Science, 1997, 45, 653-664.

Mautner, M. N.; Conner, A. J.; Killham, K.; Deamer, D. W. "Biological potential of extraterrestrial materials. 2. Microbial and plant responses to nutrients in the Murchison carbonaceous meteorite." Icarus 1997, 129, 245-257.

Mautner, M. N. "Formation, chemistry and fertility of extraterrestrial soils: Cohesion, water adsorption and surface area of carbonaceous chondrites. Prebiotic and space resource applications." Icarus 1999, 137, 178 – 195.

Mautner, M. N. and Sinaj, S. "Water-extractable and exchangeable phosphate in Martian and carbonaceous chondrite meteorites and in planetary soil analogues." Geochim. Cosmochim Acta 2002, 66, 3161-3174.

Appendix A3.3

Journal of the British Interplanetary Society, Vol. 32, Pp. 419-423. 1979

DIRECTED PANSPERMIA: A TECHNICAL AND ETHICAL EVALUATION OF SEEDING NEARBY SOLAR SYSTEMS

M. Meot-Ner (Mautner)
The Rockefeller University, New York, N.Y. 10021, USA

G. L. Matloff
Dept. of Applied Science, New York University, 26-36 Stuyvesant St., New York 10003, USA

Abstract

Advanced solar sail, deployed at appropriate locations in heliocentric space will be capable of ejecting small (payloads of moderately radiation protected biological material from the Solar System at 0.0001-0.001c. For deceleration at the target a low technology device can be provided by silvering both sides of the interstellar sail and aiming directly at a solar-type star. The probability of capture by target planets will be enhanced by the dispersion of pansperms in a circular orbit intersecting with the ecliptic plane in the target ecosphere. Biological criteria for interstellar panspermia selection include radiation resistance, and the microorganisms' nutritional requirements. Interstellar panspermia is proposed on the basis of the following ethical considerations: the moral obligation to insure the survival of the fundamental genetic heritage common to all living organisms; the desire to promote the evolutionary trend of the conquest of novel and more extensive habitats by living matter; and the pursuit of an intuitive drive to affect natural history on a universal scale. An ethical objection is the possibility of interference with indigenous biota. The chances for a destructive interference may be minimized by the proper selection of pansperms.

---

1. Introduction

The concept of interplanetary or interstellar migration of simple life forms, as a theory for the origin of terrestrial life, can be traced to

261

antiquity. A scientific evaluation of radiopanspermis, or the propagation of microorganisms by stellar radiation pressure, was first given by Arrnheis [1], but later rejected in the grounds that unprotected microooganisms would be destroyed by U.V. radiation [2]. The hypothesis of lithopanspermia, the interstellar transportation of microorganisms entrapped and protected in meteorites [3] was rejected because of the low probability for the capture of meteorites originating in other solar systems [3]. Nevertheless, the panspermia hypothesis was revived by Crick and Orgel [5] who suggested that microbial life may have been spread to new planetary habitats by design, by an early civilization extant in the galaxy at the time of the origin of life on Earth; and that the abundance of trace elements in contemporary organisms could serve as a test for the origin of life in an extraterrestrial environment. Crick and Orgel [5] did not evaluate in detail the technological requirements for engineered interstellar panspermia, though they suggested that a study of the required propulsion system would be of value. Indeed, such an evaluation is necessary if we desire to evaluate the level of sophistication that intelligent civilizations in general must achieve before panspermia by design becomes possible. This question is of general interest since engineered panspermia may constitute one mechanism for the spread of organic life in the Universe.

In this report we shall point out that a planned panspermia program designed to seed nearby solar systems with terrestrial life could be accomplished by a civilization on our contemporary level of technical sophistication. Specifically, we shall outline a strategy for engineered panspermia using current or near future technology. The design incorporates the essential elements of the radiopanspermia and lithopanspermia mechanisms; propulsion of the microbial payload will be achieved by radiation pressure using a solar sail device, and the microorganisms will be encapsulated to provide protection against U.V. radiation. An important feature of the proposed strategy is that capture of pansperms at the target solar system is not predicated on the use, and requisite interstellar survival, of complex automated navigational devices. Rather, the solar sail itself is used as a primitive deceleration device, and dispersion of pansperms in the target ecosphere serves to enhance the probability of capture.

Evidently the implementation of a panspermia project by any civilization can have no utilitarian objectives, and it will have to be motivated by philosophical considerations. The elements of motivation are not less vital for the eventual implementation of engineered panspermia than are the technical aspects. This fact was also recognized by Crick and Orgel [5], but again they did not examine this subject in detail. Because of the importance of the motivation factor, we shall survey in some detail the ethical implications of spreading, by intelligent design, terrestrial life in the universe.

2. Interstellar Launch of the Panspermia Payload

As an example of a panspermia mission we shall consider the use of a flat solar sail to launch a payload of 10kg to an interstellar trajectory. The acceleration imparted to the sail and payload by solar radiation pressure is calculated from Eq. 1

$$a = \frac{(1+P)\, A\, S\, R_e^2}{c\,(\sigma A + m_p) R^2} - \frac{G\, M_s}{R^2} \tag{1}$$

The first term on the right hand side is derived from the formula of Tsu [6] and represents the acceleration imparted by solar radiation pressure; the second term represents the solar gravitational acceleration. Here P is the reflectivity of the sail, taken as 0.9; A and σ the sail area and thickness, respectively, the latter in units of g cm^{-2}. R_e is the mean Sun-Earth distance and R is the instantaneous distance of the sail from the Sun; G the gravitational constant; M_s the solar mass; S the solar constant. The payload mass m_p will be selected so that $m_p=0.1\,(\sigma A) =0.1\, m_s$, where m_s is the sail mass. Fig 1 represents the trajectories and final velocities computed for sails of thicknesses of 10^{-5}, 10^{-4}, and 1.3×10^{-4} gcm^{-2}, launched from an orbit at 1 au.

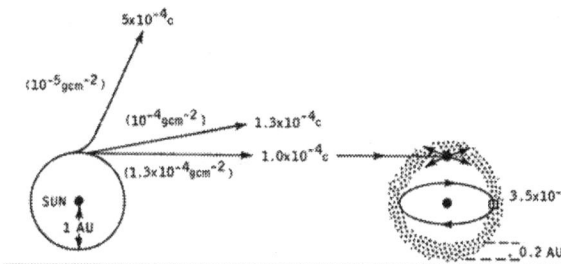

Fig. 1. Schematic illustration of the launching and capture of a solar sail device and the dispersion of the encapsulated microbial payload, for a mission whose physical parameters are discussed in the text.

Initial trajectories and final cruise velocities of vehicles, for sail thickness as indicated in parenthesis, are indicated on the left hand side diagram.

We note that films of 10^{-4} gcm^{-2} are currently available commercially. Thinner sails will be feasible to manufactory by vapour deposition techniques under micro-gravity high vacuum space conditions. A planar sail with m^s = 100kg and $\sigma = 10^{-4}$ gcm^{-2} will require a radius of 178 m, a size of feasible dimensions.

Launching of the solar sail device from a solar orbit can be achieved by prompt exposure to solar radiation, for example, by ejection of a massive shield or the closing of a "venetian blind" structure. In this article we shall consider the example of a sail of 1.3x 10^{-4} gcm^{-2}. For a sail of this thickness the two terms in the right hand side of Eq. 1 cancel, i.e., radiation pressure exactly balances gravity at any distance from the sun. Exposure to solar radiation will therefore cause the spacecraft to move in a linear outbound trajectory at the orbital velocity of 10^{-4}c. Higher speeds may be obtained by thinner sails as noted above, or by positioning the spacecraft in an orbit with a radius smaller than 1 au prior to launch. To obtain small starting orbits, the device may be maneuvered prior to the interstellar launch by orienting the sail such as to increase the orbital velocity. Trajectories aimed at targets out of the plane of the ecliptic may also be achieved by maneuvering the sail in cranking orbits prior to interstellar launch [7].

3. Navigation, Drag sand Erosion of the Interstellar Sail

To assure arrival of the spacecraft within the ecosphere of the target solar system, satisfactory navigational accuracy must be achieved. Currently, spacecraft attitudes can be monitored to 10^{-2} arc sec by interferometric star tracking [8]. Launching from 1 au, this precision allows a launch window of 0.2 sec. For a target at a distance of 10 light years, whose movement is defined exactly, this accuracy will assure entry into an orbit defined within ±0.024 au. However, the limiting factor at present time is the resolution of the measurement of stellar motion. Relative proper motions of starts at low galactic latitudes were determined recently with an average standard error of ± 0.11 arcsec/century [9]. This will result in an error of ± 120 au for a flight time of 10^5 years. The resolution of star motion may be improved by orders of magnitude in the near future by long-base-line interferometry using the forthcoming space telescope and its descendants. If the motion of the target star can not be defined more accurately, an increased number of missions can be used to assure that at least a few vehicles will be captured near the desired orbit at 1 au.

As the sail moves through interstellar medium, it will constantly strike interstellar particles, mostly hydrogen atoms and ions, at the rage of $A v \rho$ particles sec^{-1}, where A is the sail area, v is sail velocity and ρ the average interstellar hydrogen density.

If each particle striking the sail is elastically reflected, the fractional lost in sail momentum per unit time is given by

$$\Delta P_S / P_S = 2 M_H \sigma v / \sigma \qquad (2)$$

where M_H is the hydrogen mass. Taking $\rho = 1/cm^{-3}$, $v = 10^{-4}c$, $\sigma = 1.3 \times 10^{-4}$ggm cm^{-2} we obtain $\Delta P_S / P_S = 7.8 \times 10^{-4i}$ sec^{-1} for a voyage of 10^5 years, the original velocity will be reduced by 12% at the target due to interstellar drag.

Erosion is a potential problem since we desire to protect the side of the sail facing the target star so that it can be utilized during deceleration. Powell [10] has calculated that erosion will only become a problem at moderate relativistic speeds. A potential solution to both the drag and erosion problems is a sail of venetian-blind structure that will open during interstellar transit.

4. Capture of Pansperm Capsules in the Target Ecosphere

By symmetry, the trajectory of the solar sail upon approach to a sun-like target star (T_t) will be the inverse of the motion following launch. At a desired distance from S_t, e.g., 1 au, a thermally or photoactivated device can be utilized to eject the sail. The payload, moving at the orbital velocity at this point, will be captured to an orbit about S_t. This orbit will be included at a random angle to the plane of the target ecliptic, since the latter cannot be determined in advance. To ensure an acceptable probability of capture, the microbial payload will be distributed therefore in a ring which intersects the local ecliptic. For example, a payload of 1 kg containing 10^{15} microorganisms,)(each of an average mass of 10^{-12}g), will be distributed into 10^{12} capsules, each containing 10^3 microorganisms. Each of these capsules will weigh 10^{-9} g, and will be shielded by a protective film against uv radiation. A reflective coating can also be used to keep the encapsulated microorganisms at a low temperature.

A chemical propellant may be used to impart to the capsules a distribution of velocities of $\pm 5 \times 10^{-6}c$ about the mean orbital velocity of

10^{-4}c. Using the standard equations of propulsion we find that the required mass of an efficient propellant (specific impulse = 400 sec) is 0.46kg; the required propellant may thus easily be accommodated in the payload mass.

The capsules distributed in this manner will disperse in 20 years into a ring with a density of 3.5 x 10^{-15} capsules/cm^2. The pansperm ring will be spread in width from 0.90 to 1.1 au. A planet with a gravitational radius of 10^4km, whose orbit in the target intersects the microbial ring, will thus capture 1.1x10^4 capsules, or 3.3 x 10^7 microorganisms per passage through the ring, as illustrated in Fig.1.

Capsules consisting of 10^3microorganisms will have a radius of about 7 microns. Their orbit will not be affected substantially by solar radiation pressure (Arrhenius, quoted by Tobias and Todd [11]) and they will stay in their orbit indefinitely. Multiple passage of the target planet through the microbial ring will thus be possible.

With proper adjustment of capsule size and reflectivity, a predetermined portion of the capsules may be designed to be swept outwards gradually by radiation pressure, transit the cross-section of the target ecosphere and seed all the planets therein. An interesting further elaboration may be implemented if the reflectivity of the capsule coating material is rendered inversely variable with incident radiation. I this case the capsule parameters may be so adjusted that they will oscillate between orbits of preset minimum and maximum radii, corresponding to the limits of the ecosphere. In this case, all the microbial capsules will be swept up by planetary or asteroid objects I this volume.

Since the relative orientations of the capsule and target planetary orbits are random, high relative velocities (up to twice the orbital velocity) may lead to the burnup of the capsules in a substantial fraction of the missions. At the other extreme, however, when the orbital orientations of the capsules and the target accidentally coincide, small relative orbital velocities of the capsules should assure safe entry to the target atmosphere. Once landed, the thin metal or organic foil coating of the capsules should corrode with ease in the planetary atmosphere or hydrosphere, and free distribution of the microorganisms into the planetary environment will be permitted.

5. Biological Considerations

The basic biological criterion is that the pansperm payload is required to satisfy survival during transport and landing, and growth in the target environment. The question of the survival of microorganisms in space and under various planetary conditions have been the subject of a great volume of investigation in relation to planetary quarantine [12] and we shall address briefly only a few related points.

The major source of lethal damage to microorganisms in space is uv radiation. However, protection against uv damage may be provided by very thin shielding. For example, a chromium film of 800 angstrom is sufficient to protect microorganisms against solar uv [13]. A shielding metal film of this thickness for a capsule of 10^3 microorganisms, weighing 10^{-9}g, will weigh only 7.3 x 10^{-11} g. Thus, uv shielding will not add substantially to the payload mass.

On the other hand, shielding against ionizing radiation requires a thickness which is prohibitive in the present context, although the explosive propellant used for capsule dispersion and other structural elements may provide some protection during interstellar transit. Microorganisms that survive doses in the range of 10^6 rad are known [11]. Less resistant strains can adapt to radiation; low-temperature, frozen and dehydrated conditions, which may be achieved under interstellar transit, can also increase resistance up to an order of magnitude. We thus consider 10^6 rad as the limiting permissible does of ionizing radiation. Given an average flux of galactic ionizing radiation of 10 rad/year [14], a voyage of 10^5 year is permissible. Minimum shielding and low temperatures may allow an interstellar transit time of 10^6 years. This time scale is adequate to reach targets at distances of 1-=100 light years by solar sail propulsion.

To grow at the target planet, the selected pansperms will almost certainly have to be anaerobic. It is also desired that the pansperms will be photosynthetic, autotrophic and/or spore forming. The first two traits will facilitate independence in terms of energy and nutrition. The use of spores will enhance the probability of survival under extreme conditions.

Averner and co-workers [15] recently studied the habitability of Mars, and concluded that blue-green algae, or probably a strain combining desired characteristics of several species of algae, may well be able to grow under Martian conditions. This also appears a suitable choice for panspermia candidates. Other species adapted to survive in reducing environments may also be included in the payload. For example, the autotrophic, anaerobic, photosynthetic species Thiospirillum and Chromatium may be suitable for early terrestrial type environments where CO_2 and H_2S both may be available. The large number of microorganisms in a mission will allow combining in each capsule species suited to grow under diverse conditions.

6. Ethical Motivation

The ethical arguments that may induce us to implement panspermia missions or to refrain from doing so are necessarily subjective and qualitative. The ethical arguments that we shall survey are based on notions prevalent in contemporary natural philosophy.

a. *Promoting the evolutionary trend to the conquest of new habitats.*
    The primary rationale for engineered panspermia, in our view, is to *promote and perpetuate the genetic heritage common to all terrestrial life.* This proposition is predicated on the notion that qualities universal to all terrestrial life do exist. In the framework of contemporary biochemistry, these unifying qualities may be identified as those patterns of the hereditary and metabolic mechanisms which are shared by all cellular organisms from prokaryotes to man. These fundamental universals of terrestrial life were preserved throughout evolution, and adapted to function in diverse external environments, including extreme conditions of pH, temperature, pressure and water activity. Indeed, the trend to conquer all accessible habitats may be seen as a characteristic manifestation of biological evolution.

    In this framework, we may propose that the unique human capacities of cognition and manipulation imply a moral obligation on the part of our species towards the totality of terrestrial life. This obligation suggests that technology should be used to promote the conquest of new habitats by living matter, as extensively as our technology permits. Evidently, engineered panspermia will serve this object.

b. *Interference with indigenous biota*
    The possibility that pansperms will disrupt or destroy an indigenous biosystem presents the most severe argument against engineered panspermia. The possibility of destructive interference with indigenous organisms could be eliminated only by the close range survey of the target; this is far beyond the level of technology considered here. However, we wish to present some arguments which may mitigate this problem.

    The most tragic outcome of a panspermia mission would be to harm intelligent organisms endowed with self-awareness. This would happen only if these inhabitants have not yet developed the means to eliminate primitive biological invaders. It is likely, however, that the evolution from the inception of consciousness to a complete biological control of the environment should be in general rapid, occupying at most a few million years, as seems to be likely in our case. To encounter a civilization at this brief phase of its evolution, in the near vicinity of our solar systems, must be highly improbably by any estimate.

More generally, to cause damage to indigenous organisms by direct infection, their biology must be essentially similar to our own: for example, silicon-based organisms, or even ones using D-amino acids, will be probably unharmed. Interference by competition for food or energy resources is also possible. However, such competition may equally well arise from new species evolving by natural ways in the local biosphere. Moreover, harm to an ecosystem which have evolved photosynthesis may be avoided by selecting as pansperms strictly anaerobic organisms which will be destroyed by exposure to the oxygen atmosphere upon arrival. Thus the nature of the organisms which could be harmed by the pansperms may be confined into relatively narrow limits.

If competition between the pansperms and the local biota for vital resources should commence, it will but constitute an interstellar extension of the evolutionary struggle for the survival of the fittest. In this instance, panspermia will have served as a vehicle for the cosmic extension of organic evolution. This eventuality is not necessarily more evil than natural evolution on a local scale.

c. *Further Motivations: A Genetic "Noah's Ark"; Manipulation of Natural History; Cosmic Loneliness*
Engineered panspermia may become especially urgent if a catastrophic development threatening all life, or at last mankind of human civilization, appears imminent. A life system threatened with extinction, or one which perceives itself as such, may desire to perpetuate its genetic heritage by transplanting it to new habitats. Such motives were suggested by Crick and Orgel [5], Sagan and Shkloviskii [16] and others. A threat to the survival of our technological civilization may also suffice to motivate a panspermia project, since advanced technology will be required to escape the ultimate incineration of all terrestrial life by the Sun at its red giant phase.

It is generally recognized that current technology has the capacity to effect a global catastrophe; although it is impossible to estimate the actual probability that such a catastrophe will occur. However, the desirability of a panspermia project will be decided upon not on the basis of the actual, but the *perceived* probability that our civilization, human survival, or life on Earth in general is threatened. The survival time of the current civilization has been estimated in context of the search for extraterrestrial intelligence by several authors [17]. In general, the lifetime of this civilization is perceived as limited, estimates ranging from tens to thousands of years. For example, von Hoerner [18] estimates that an advanced civilization will destroy all higher life on its planet within 30 years with a probability of 60%, and that a catastrophic nuclear war will become more probable than peace after only 45.6 years even if the probability for war in every seven year period is only 10% [19]. Such expectations seem to be widespread; for example, an informal poll by Westman [20] showed that students estimate the longevity of our civilization as 100-200 years. We also conducted an informal survey of 32 young, mostly college-educated subjects. The average perceived probability for the destruction of all life within 500-1000 years through a man-made disaster was 22%, and for the destruction of civilization, 42%.

It is hard to assess the level of a perceived threat that would suffice to motivate our civilization to engage in directed panspermia; it would be even harder to assess the reactions of a past extraterrestrial civilization to a perceived threat. Nevertheless, it is interesting to note that in our own civilization the emergence of the technological level which makes panspermia possible generates, simultaneously, a threat that may also make directed panspermia desirable [21].

A further possible motivation for engineered panspermia may be provided by the expectation that it will afford a profound influencing of natural history by human design. Beyond transforming the history of the target ecosphere, the descendants of pansperm evolution - if not our own descendants - may in the long run further spread life in the universe. Via panspermia we may thus ultimately contribute to turning biological activity into a determinant force in the physical evolution of the universe.

Finally, the growing perception of the magnitude of the cosmos, and the absence of evidence for extraterrestrial life so far tend to induce a growing sense of our cosmic isolation. White the search for extraterrestrial life may lead to a passive solution, engineered panspermia will provide an active route of escape from the stark implications of cosmic loneliness.

7. Conclusion

We presented a scheme for engineered panspermia which relies only on propulsion by radiation pressure and other current or near future technologies. The point is made that in our own civilization these means were developed within decades after the advent of advanced technology. Also we noted that in a civilization with our level of general intelligence, ethical motivations for engineered panspermia arise concurrently with the advent of the required technology. All this indicates that there exist no significant barriers, technological or psychological, to the implementation of engineered panspermia by a civilization even at the earliest stages of its technological phase. Engineered panspermia could thus constitute a facile avenue for the spread of organic life in the galaxy.

References

1. S. Arrhenium, "Veldarnas Ultveckling" Stockholm, 1908, Cited by A.I. Oparin in "Genesis and Evolutionary Development of Life", Academic Press, New York 1906.
2. A.D. McLaren and D. Shugar, "Photochemistry of Proteins and Nucleic Acids", Macmillan, New York (1964).
3. C.A. Tobias and P. Todd, Radiation and Molecular and Biological Evolution. In "Space Radiation Biology and Related Topics", C.A. Tobias and P. Todd, eds., Academic Press, New York , pp197. A brief review of panspermia is given in this reference (1974).
4. C. Sagan, Private communication, quoted in Crick and Orgel, (1973).
5. F.H. Crick and L.E. Orgel, Directed Panspermia, *Icarus,* **19**, 341 (1973).
6. T. C. Tsu, Interplanetary Travel by Solar Sail, *ARS Journal* pp.422 (1959).
7. L. D. Friedman, Halley Rendezvous via Solar Sailing-Mission Description, *Bull. Am. Astronom. Soc.,* **9**, 463, (1977).
8. A. B. Decon, Interferometric Star Tracking, *Applied Optics,* **13**, 414 (1974).
9. R. D. Stone, Mean secular Parallax at Low Galactic Latitude, *The Astronomical Journal.,* b **38**, 393 (1978).
10. C. Powell, Heating and Frag at Relativistic Speeds, *J. Brit. Interplanetary Soc.,* **28**, 546 (1975).
11. C. A. Tobias and P. Todd, Cellular Radiation Biology. In "Space Radiation Biology and Related Topics", C. A. Tobias and P. Todd, eds., Academic Press, New York, p. 162 (1974).

12. L. B. Hall Planetary Quarantine: Principles, Methods and Problems. In "Foundations of Space Biology and Medicine", M. Calvin and O. G. Gazenko, eds., NASA, vol. 1 pp. 403 (1975).

13. A. A. Imshenetskii, S. S. Abizov, G. T. Verenov, L. A. Kuzurina, S. V. Lysenko, G. G. Stonikov and R. I. Fedorova, Exobiolgy and the Effects of Physical Factors on Microorganisms, In "Life Sciences and Space Research", A. H. Brown and F. G. Favorite, eds Vol.5 pp. 250, Amsterdam, North-Holland (1967).

14. S. B. Curtis, Radiation Physics and Evaluation of Current Hazards. In "Space Radiation Biology and Related Topics", C. A. Tobias and P. Todd, eds., Academic Press, New York pp.22 (1973).

15. M. M. Averner and M. MacElroy, (Eds) "On the Habitability of Mars. An Approach to Planetary Ecosynthesis", NASA Scientific and Technical Information Service, Washington, DC (1976).

16. C. Saganand I.S. Shloviskii, "Intelligent Life in the Universe", Dell Publishing co., New York (1966).

17. For a review see C. Ponnamperuma and a. G. W. Cameron, " Interstellar Communication Scientific Perspectives", Haughton Miffin, Boston, (1974).

18. S. von Hoerner, The Search for Signals from Other Civilizations, *Science*, **134**, 1839 (1961).

19. S. von Hoerner, Populations Explosion and Interstellar Expansion, *J. Brit. Interplanetary Soc.,* **28**, 691 (1975).

20. W. E. Westman, Doomsdays Expectations, *Science,***193**, 720 (1976).

21. The fact that we ourselves are favourable inclined to directed panspermia is in itself proof that such a project can in fact be found desirable in the early stages of technical civilization.

Appendix A3.4

Journal of The British Interplanetary Society, Vol 48 pp.435-440, 1995

DIRECTED PANSPERMIA. 2 TECHNOLOGICAL ADVANCES TOWARD SEEDING OTHER SOLAR SYSTEMS, AND THE FOUNDATION OF PANBIOTIC ETHICS

MICHAEL NOAH MAUTNER
*Department of Chemistry; University of Canterbury, Christchurch 8001, New Zealand

Department of Chemistry, Virginia Commonwealth University, Richmond, Virginia 23284-2006, USA

Abstract

The technologies required for seeding other solar systems with terrestrial life are undergoing rapid progress. Some key developments are: for navigation, accurate astrometry toward resolution of 0.1 milliarc-seconds; for facile propulsion and braking methods, advances in solar sailing; for payload selection and development, research on extremophile microorganisms and PCR methods of guided evolution. The identification of young planetary systems afford targets without interference with local biota avoiding a major ethical obstacle, and provides long-lived habitats for higher evolution. Concurrent with the technology, a panbiotic ethical world-view is also emerging that identifies propagating Life in the universe as the ultimate human purpose. In a few decades our civilization can posses both the means and motivation to initiate evolution in other solar systems.

Appendix A3.5

Journal of The British Interplanetary Society, Vol. 50, pp. 93-102, 1997

DIRECTED PANSPERMIA. 3. STRATEGIES AND MOTIVATION FOR SEEDING STAR-FORMING CLOUDS

MICHAEL N. MAUTNER
Department of Chemistry, Virginia Commonwealth University, Richmond, VA 23284-2006 USA and Lincoln University, Lincoln, New Zealand* (MautnerM@Lincoln.ac.nz)

Abstract

Microbial swarms aimed at star-forming regions of interstellar clouds can seed stellar associations of 10 - 100 young planetary systems. Swarms of millimeter size, milligram packets can be launched by 35 cm solar sails at 5E-4 c, to penetrate interstellar clouds. Selective capture in high-density planetary accretion zones of densities > 1E-17 kg m-3 is achieved by viscous drag. Strategies are evaluated to seed dense cloud cores, or individual protostellar condensations, accretion disks or young planets therein. Targeting the Ophiuchus cloud is described as a model system. The biological content, dispersed in 30 μm, 1E-10 kg capsules of 1E6 freeze-dried microorganisms each, may be captured by new planets or delivered to planets after incorporation first into carbonaceous asteroids and comets. These objects, as modeled by meteorite materials, contain biologically available organic and mineral nutrients that are shown to sustain microbial growth. The program may be driven by panbiotic ethics, predicated on:

1. The unique position of complex organic life amongst the structures of Nature;
2. Self-propagation as the basic propensity of the living pattern;
3. The biophysical unity humans with of the organic, DNA/protein family of life; and
4. Consequently, the primary human purpose to safeguard and propagate our organic life form.

To promote this purpose, panspermia missions with diverse biological payloads will maximize survival at the targets and induce evolutionary pressures. In particular, eukaryotes and simple multicellular organisms in the payload will accelerate higher evolution. Based on the geometries and masses of star-forming regions, the 1E24 kg carbon resources of one solar system, applied during its 5E9 yr lifespan, can seed all newly forming planetary systems in the galaxy.

Introduction

Panspermia, natural or directed, is a possible mechanism for the spread of life through interstellar space [1-7]. In fact, we may be already capable to use solar sail technology for seeding nearby new planetary systems with our DNA/protein form of life [4-6]. The program can become realistic in decades, due to rapid advances in high-precision astrometry, advanced propulsion, discovery of extrasolar planetary systems, and microbial genetic engineering [5]. An essential component for realizing directed panspermia is the ethical motivation. Seeding distant planets with life is the ultimate altruism, bearing results long after the generations that implement it. The ethical motivation for such a program must recognize

(1) the unique position of complex, self-propagating organic Life in Nature;

(2) the unity of all organic, cellular DNA/protein life, from microbes to humans and post-humans;

(3) and, consequently, the primary human purpose, to safeguard and propagate our life-form [4,5].

Prime targets for biological expansion can be regions of interstellar clouds where newly forming stars and planetary systems are concentrated. The discussion below will consider the physical environments of such regions, and the implications for the microbial missions. The article will survey both the technological and ethical aspects of seeding with life star-forming interstellar clouds.

Target Environments: Star-Forming Clouds, Dense Cores, and Planetary Accretion Disks

The mission will be illustrated by choosing a representative candidate, Rho Ophiuchus (distance = 520 ly), a cloud that forms long-lived low and medium mass stars. As described by Mezger [8], the overall cloud extends to about 50 ly as low density gas (hydrogen atom density $nH < 1E3$ cm-3, (i.e., $< 1.7E-18$ kg m-3)) of total mass » 3,000 M¤ (solar mass M¤ = 2E30 kg), and contains a 6x6 ly dense fragment with a density of 1E4 cm-3 and mass ~500 M¤ , containing 78 young stellar objects of low-mass dust-embedded or early accretion stage T Tauri stars. Within this cloud are four cores with diameters of » 1 ly and, densities of densities of 1E6 cm-3 (1.7E-15 kg m-3) and masses of 1 - 15 M¤ .One of these cores shows four protostellar condensations with radii of » 3E14 m, densities of 1E7 cm-3 (1.7E-14 kg m3) and masses of 0.4 to 3 times the mass of the sun. Dust temperatures in this region are 15 - 20 K.

Small panspermia capsules captured in a protostellar condensation or about a young star in an accreting planetary system will become part of the dust in the system. The protostellar condensation free-falls in » 4E4 yr to cores with radii of 100 au and densities of 1E11 - 1E12 cm-3 (1.7E-10 - 1.7E-9 kg m-3), which collapses further during 1E5 - 1E6 yr into a 1E6 m thick, 100 au (about 1E13 m) radius dust ring [9], that comprises 0.01 M¤ (2E28 kg) (possibly up to 0.1 M¤ (2E29 kg)) mass about a 1 M¤ young T-Tari star, and has a temperature of T = 50 - 400 K at 1 au (consider 250 K) (with possible periodic heating over 1,000 K), and T = 250a-0.58 at other distances a (in au = 1.5E11 m units) [10]. In the ring, the dust accretes rapidly (in 1E3 - 1E4 periods of revolution) from micron-size grains to 1 - 10 km planetisimals; then, in about 1E5 years, to 1E3km radius, 1E21 kg runaway planetary seeds that developing into 1E23 kg planetoids; and in the next 1E8 years, to planets [10]. Most of the gas is ejected from the disk in 1E6 - 1E7 yr by bipolar outflow and stellar UV radiation [10]. A fraction of the residual materials accrete in a zone of several tens of au from the star to become 10 km diameter, 1E14 - 5E14 kg nuclei of 1E13 comets, most of which are expelled to interstellar space [11], except 1E11 - 1E12 comets with a total mass of 1E25 - 1E26 kg that are retained in the Oort cloud at 1.7E4 - 1E5 au. [12] Another about 1E23 kg materials form the Kuiper belt comets [13], and 1E22 form the main-belt asteroids [14].

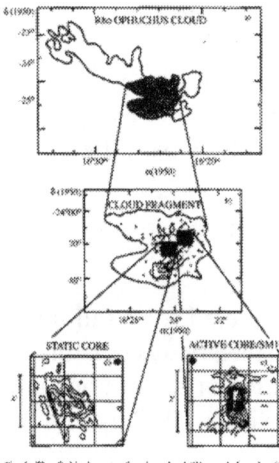

Figure 1. The Rho Ophiuchus cloud, a potential target for directed panspermia. See enlarged picture in Chapter 3 "Seeding the Universe".

a. Low-density envelope or mass 3000 M_{SUN} is shown shaded.
b. The dense region, showing the positions of 78 young stellar objects.
c. Active star-forming core of mass 15 M_{SUN}. d. Static, non star-forming core. (From P.G. Mezger, "The Search for Protostars Using Millimeter/Submillimeter Dust Emission as a Tracer", in "Planetary Systems: Formation, Evolution and Detection", B.F. Burke, J.H. Rahe and E. E. Roettger, eds., Fig 6, p. 208. Reproduced with kind permission from Kluwer Academic Publishers).

Cometary mass ablating in transits maintains a Zodiacal dust cloud of 2.5E16 kg and mean lifetime of 1E5 yr by injecting at present, about 2E4 kg s-1 dust near the perihelion passes at <3.5 au [15]. Of this, 0.15 kg s-1, i.e., a fraction of 1E-5, is collected by the Earth [16a]. At higher densities in the prebiotic period between 3 and 4 Gyr (1Gyr = 1E9 yr) ago, 1E17 kg of the cometary dust accreted onto the Earth in the form of 0.6 to 60 m m radius particles in which organic material can be preserved during atmospheric transit [2]. Similar to the Zodiacal dust collection efficiency, 1E-5 of the asteroid fragments produced by collisions eventually impacts on the Earth as meteorites [16b]. Both data suggest that 1E-5 of the objects in orbit within several au of a habitable planet are eventually collected.

Altogether, the 1E17 kg material of cometary origin that was collected by the Earth in the early biotic period between 3 - 4 Gyr ago, constitutes about 1E-13 of the total 1 M¤ (2E30 kg) protostellar condensation, 1E-11 of the mass of the original accretion dust ring, and 1E-9 of the total present Oort cloud cometary mass.

These data from our solar systems are used as models. These data are current, model-dependent estimates with uncertainties up to an order of magnitude, and respective figures may be of course different in other solar systems.

An Overview of The Swarm Strategy

In the previous papers [4-6], we considered solar sail missions of a few vehicles targeted at specific nearby planetary systems that possess protoplanetary dust rings, such as Vega, beta Pictoris, and Fomalhout. For such missions, suitable targets should be within <100 ly for targeting accuracy, and have observable accretion disks or planets, preferably about young F, G or K type stars that will stay on the main sequence for >1E9 years to allow higher evolution. Only a few suitable objects are known.

It may be more efficient therefore to aim for nearby star-forming regions with large concentrations of accreting planetary systems. Such regions are found in collapsing dense molecular clouds that fragment to form stellar associations, some with up to 100 new 0.5 - 5 times the mass of the sun, which are long-lived stars.

The nearest suitable star-forming zones are dense regions (>106 cm-3), that are >100 ly away. It is not possible to target a few vehicles accurately at individual stars at such distances, and even if targeted, the vehicles may be scattered by the high density medium. For such environments, a statistical swarm strategy may be preferred.

The swarm strategy uses solar sails to launch large numbers of small, milligram size, microbial packets. The size of the packets is designed so that they transit the thinner cloud regions and are captured in high-density protostellar condensations, where they will fragment into small, e.g., 30 m m radius capsules. Some capsules will land on already accreted planets, while other capsules that arrive in actively accreting protoplanetary systems, will be captured in asteroids and comets. Subsequently, when host comets warm up near perihelion passages, the microbial payload in them may multiply [17]; in any event, microbes or capsules will be ejected with the cometary dust particles and like them, a fraction will be captured by planets. Alternatively, the capsules can be transported to planets when the host asteroids and comets, or their meteorite fragments, impact. Using nutrients provided in the capsule, supplemented by the rich nutrients in the host carbonaceous meteorite or cometary matrix [18,19], and subject to wet and warm planetary conditions, the microbial payload can then start to multiply. Materials from the planet will mix with the capsule and meteorite microenvironments, and the microorganisms can adapt gradually to the planetary chemistry. Finally, the microorganisms will break free to multiply and evolve in the environment of the new planet.

This sequence will be evaluated below quantitatively, to estimate the probability of success and the required amounts of panspermia material.

3.1 Propulsion and Launch

Our previous papers considered technologies for sending large microbial payloads on the order of 10 kg to nearby solar systems [4-6]. We considered relatively simple technology, using solar sail vehicles with areal densities 1E-4 kg/m2 with thin sails of thickness 1E-7 m (0.1 microns), and of sizes on the order of 1E6 m2, which can reach velocities of 5E-4 c when launched from 1 au. The sails must remain stable during transit times of 2E5 years to targets up to 100 ly away, so that they can provide braking by radiation pressure after arrival.

In comparison with the 10 kg payloads of directed missions, the swarm approach launches large numbers of small payloads. The considerations below suggest launching 1 mm radius, 4.2E-6 kg microbial packets. Therefore, the swarm method miniaturizes the mass of each launched payload by about a factor 2E6, which further reduces the technological requirements and may allow new propulsion approaches. Once in the target region, the packets can further decompose into 4E4 capsules of 30 m radius containing 1.14E-10 kg microbial mass, that is appropriate for eventual non-destructive atmospheric entry. The large numbers can also increase the probability of capture.

Even for the milligram payloads, the most imminent technology appears to be solar sailing. For effective devices, the sail/payload ratio should be about 10:1, requiring sails of 4.2E-5 kg. With an areal density of 1E-4 kg m-2, this will require sails of 0.42 m2, i.e., sails with a radius of 0.35 m. Such small sails can be mass manufactured easily, which is important since very large numbers are required. For planetary targets in the dilute medium within 100 ly, the 30 T m, 1.1E-10 kg capsules can be launched individually, using 1E-9 kg sails of 0.18 cm radius. These miniature objects can be mass manufactured and launched even more easily.

The thin sail devices with σ_a = 1E-4 kg m-3 could transit the local low-density medium about the Sun with little drag. However, the sail devices cannot penetrate even a diffuse interstellar cloud with r m = 1E-19 kg m-3, where they will stop rapidly, for example, slow down to 15 m s-1 in the first 0.4 ly. For this reason, and to minimize scattering during transit, a useful strategy would be for the sails to eject the capsules once they obtained the final velocity of 1.5E5 m s-1, possibly with an impulsive ejection using the sail as countermass, to impart the payload further acceleration. Alternatively, the sails may be manufactured of biopolymers that would fold over the payload after exit from the solar system. They can then provide additional shielding in transit, and be used as a nutrient shell once the capsules land on the host planet.

The transit time for a sail-launched capsule to a cloud 100 ly away is 2E5 years, during which the payload will be subject to 2E6 rad of ionizing radiation. This can be lethal, or at least strongly damaging to most microorganisms. It may be desirable therefore to use alternative propulsion methods to achieve greater velocities and shorter transit times. However, at high speeds, ablation and heating of the capsules can be significant, especially in the dense cloud area, requiring velocities <0.01 c. At such high entry velocities, even sub-millimeter size, sub-milligram capsules may penetrate the clouds sufficiently, so further miniaturization of the microbial packets down to microgram levels may be possible.

3.2 Astrometry and Targeting

The large size of star-forming regions, compared with individual planetary systems, is a major advantage. Compared with astrometry requirements for targeting a habitable zone about a specific star, on the order of several au (1E11 - 1E12 m), the size of the model star-forming Ophiuchus cloud fragment is larger by a factor of 10,000 i.e., about 6 ly (6E16 m). In terms of angular resolution, a 1 au planetary target zone at 50 ly distends 1.8E-5 degrees, whereas the 6 ly Ophiuchus fragment at 520 ly distends 0.68 degrees as seen from Earth.

Given the substantial space velocities of interstellar clouds, on the order of 1E4 m s-1, the vehicles must be aimed at the expected position of the targets at the time of arrival. The uncertainty in calculating this position arises from the limits of the resolution of the proper motion of the cloud when the vehicles are launched. The positional uncertainty at the time of arrival, δy, is expressed by equation (1), where α_p is the resolution of proper motion, d is the distance from Earth, and v the velocity of the vehicle (α_p in arcsecs/yr, other units in SIU).

$$\delta y = 1.5E\text{-}13\ \alpha_p\ (d^2/v) \tag{1}$$

Angular proper motion resolutions of 1E-5 arcsec/yr can be anticipated. The positional uncertainties of the various targets considered upon the arrival of fast (v = 0.01 c) or solar sail based (v = 1E-5 c) missions, i.e., the L y values, are listed in Table 1. Note that for the large cloud core, and even for individual protostellar condensations, the uncertainty is smaller than the radius of these objects.

Given the uncertainty δy in the position of the target when the vehicles arrive, the panspermia objects should be launched with a scatter, to arrive in a circle with radius δy about the calculated position (scatter with a Gaussian distribution may be more effective). The probability that the vehicle will actually arrive in the target region, P_{target}, is then estimated from the ratio of cross-sectional areas of the target region to that of the area of the targeting scatter. Equation (6) in reference [5] was derived on this basis, and similarly we obtain equation (2) for a spherical target with a radius r_{cloud} with cross-sectional area $A_{target} = \pi r^2$. For planetary targets within a habitable zone of radius R_{hz} and width $w_{hz} = 0.4_{hz}$, the area of the target habitable zone is equal to that of a circle with radius $r = 0.89\ r_{hz}$.

$$P_{(target)} = A_{(target)} / \pi\ (\delta y)^2 = 4.4E25\ r_{target}^2\ v^2 /\ \alpha_p d^4 \tag{2}$$

For cases where the area of the target is larger than of the positional uncertainty, we obtain $P_{target} > 1$, which may be interpreted as approximately unit probability. Equation (2) yields the P_{target} values as shown in Table 1. Note that most of the microbial packets will arrive in the targeted star-forming cloud region, and even the smaller specific protostellar condensations can be targeted accurately. In fact, even with a reduced resolution of 1E-4 arcsec/yr, the dense core can be targeted reliably. However, even with α_p of 1E-5 arcsec/yr, P_{target} for a 100 au radius dust sphere about a dust-embedded star or accretion disk (perpendicular to the Earth-star axis) is 3.6E-3, and for 1 au habitable zone about a star at the same distance of 520 ly is only 3.9E-7. Targeting these smaller specific objects at these distances is inaccurate because of the d^{-4} dependence of the P_{target} function.

3.3 Capture at the Target Zone, and Considerations of Capsule Size

In the target interstellar clouds, the density increases gradually from the diffuse cloud to a dark cloud fragment, dense cores, protostellar condensations and accretion disks. This allows designing the capsule geometry (size) for selective capture in the desired zone, based on drag by the medium as given by equation (3) for elastic collisions with gas molecules [6].

$$dv/dt = -2(\rho_m v^2 A_c/m_c) \qquad (3)$$

Here ρ_m is the density of the medium; v is the velocity, A_c the area and m_c the mass of the capsule. Note that $A_c/m_c = 1/\sigma_a$, where s(a) is the areal density of the object. For a spherical object, $\sigma_a = (4/3)\rho_c r_c$, where σ_a is the density of the capsule material, assumed to be 1E3 kg m^{-3} for a biological payload, and r_c is radius of the capsule. Using these relations we can substitute for A_c/m_c in equation (3) to give the radius directly as a variable in equation (4), which was used for numerical integration.

$$dv/dt = -(3v^2/2\rho_c) \rho_m/r_c \qquad (4)$$

In these calculations we consider spherical capsules entering the cloud with a velocity of 1.5E5 m s^{-1}, and consider that their velocity becomes homogenized with the medium when they are decelerated to 2E3 m s^{-1}, a typical internal velocity of grains in a cloud. Since most of the distance is covered during the high velocity entry period, continuing travel under further deceleration has little effect on the depth of penetration. Calculations also show that acceleration due to the gravity of the cloud adds only an insignificant velocity increment of about 1E4 m s^{-1} before entry to the cloud. Other effects such as the complex gravitational and magnetic fields in the clouds require further study. Note that in equation (4) the critical variable is ρ_m/r_c, i.e., for a given desired penetration depth, the capsule radius has to vary proportionally with the density of the medium.

To reach the dense protostellar regions or accretion disks, the microbial packets need to penetrate first through the less dense, but larger zones in the diffuse cloud, the dark cloud fragment and the dense core. A figure in Part 4 Panspermia shows the deceleration of spherical objects with radii of 35 μm and 1 mm, injected into these clouds with an initial velocity of 5E-4 c (1.5E5 ms-1), in terms of velocity vs. penetration distance as computed using equation (4), along with the radii of the various zones. Note that both size objects penetrate fully the Ophiuchus cloud fragment. The 35 μm object is stopped in the dense core, but the 1 mm object can penetrate it to the even denser protostellar condensations, where both objects are stopped well before full penetration. In this region, the 1 mm object penetrates only to about 0.5 of the radius. This is adequate so that the capsule will be incorporated into the dust cloud. In fact, larger objects with r > 35 mm would transit the protostellar region and would not be captured. These calculations illustrate the use of microbial capsule size for selective capture in desired regions.

Table 1. Parameters for advanced (v = 0.01 c) and solar-sail (v = 5E-4 c) microbial swarm missions to nearby stars and to the Rho Ophiuchus cloud.

| | d (ly)[a] | r (m)[b] | dy (m)[c] v (c) | |
			0.01	5E-4
Nearby Stars				
Alpha PsA	22.6	5.0E11	2.3E10	.7E11
Beta Pic	52.8	1.3E12	1.3E11	.6E12
Rho Ophiuchus Cloud				
Dense Core	520	3E16	1.2E13	.5E14
Protostellar Condensation	520	3E14	"	"
Early Accretion Disk	520	1.5E13	"	"
Late Accretion Disk	520	2.6E12	"	"
Young Stellar Object	520	5.3E11	"	"

Table 1 (continued).

	P_{target}[d] v (c)		P_{planet}[e] v (c)		m (kg)[f] v (c)	
	0.01	5E-4	0.01	5E-4	0.01	5E-4
Nearby Stars						
Alpha PsA	(4.7E2)	(1.2E0)	1E-5	1.1E-5	1.1E-3	1.1E-3
Beta Pic	(1.0E2)	2.5E-1	1E-5	2.5E-6	1.1E-3	4.5E-3
Rho Ophiuchus cloud:						
Dense Core	(6.3E6)	(1.4E4)	1E-16	1E-16	1.1E8[g]	1.1E8[g]
Protostellar Condensation	(6.3E2)	(1.4E0)	1E-13	1E-13	1.1E5	1.1E5
Early Accretion Disk	(1.6E0)	3.6E-3	1E-11	3.6E-14	1.1E3	3.1E5
Late Accretion Disk	4.7E-2	1.1E-4	1.5E-11	3.5E-14	7.5E2	3.2E5
Young Stellar Object	1.9E-3	4.4E-6	1.9E-8	4.4E-11	5.9E-1	2.6E2

a. Distance to the target.

b. Radius of the target objects. For planets, the radius of a circle with an area equal to that of the habitable zone i.e., $r = 0.89r_{hz}$. alpha PsA and Beta Pic, r_{hz} from ref. 5, for 1 solar mass young stellar object, $r_{hz} = 1$ au. For the late accretion disk, radius of a circle with an area equal to a disk from 10 to 20 au.

c. Uncertainty in target position at arrival, from equation (1).

d Probability of arrival within the target zone, from $r^2/(dy)^2$. For values of $P > 1$, shown in parentheses, the arrival probability is approximately unity.

e. Probability of capture by a planet in the habitable zone, obtained from $P_{target} \times P_{capture}$. For targeted planets, $P_{capture} = 1E-5$; for other targeted objects, see text.

f. Biomass to be launched for the capture of 100 capsules of 1.1E-10 kg at the target planet, calculated from $1.1E-8/P_{planet}$.

g. Mass requirements per planetary system. Note this mission requires launching 100 times the given masses, for distribution through the cloud.

Capture in accretion disks requires special considerations. Statistically, most objects will encounter the 1E6 m thick, 1E13 m radius disks on the disk face (rather than the edge). An early accretion disk containing the original 100:1 gas/dust ratio can be considered as a homogenous gas medium with a density (from the mass/volume ratio) of 2.8E-5 kg m^{-3}. The 1E-3 m capsule entering with v = 1.5E5 ms^{-1} will be captured at a depth of 1E5 m, about 1/10th of the thickness. At later stages of accretion, the disk becomes thinner, and dominated by increasingly large solid aggregates. Also, because of the close approach of 1E6 meter to the central plane of the disk before drag braking starts, the approaching objects may be significantly accelerated by the disk's gravity. Once the disk is gas-free, the capsules will be captured into the disk by collisions with solids, or will be captured gravitationally into circumstellar orbits. In fact, capture at the later stages of cometary accretion, into the outer cometary crust is desirable as this facilitates the subsequent release and delivery to planets.

Finally, for planetary targets, for objects placed in orbits near the planet at <3.5 au, a fraction of 1E-5 will be captured by the planet as noted above (note that this factor was not considered in reference [5]).

For maximizing the probability of success, it is desirable to maximize the number of survivable units for a given total payload mass and therefore to minimize the capsule size. From the drag considerations, the optimal size for penetrating the cloud is 1 mm. However, once in the target region, sufficient drag is in fact necessary for capture, and the capsule size can be reduced further. In fact, it is estimated that only dust particles in the range r = 0.6 - 60 μm can survive atmospheric entry and still remain cold enough to preserve organic matter [20]. A median size in this range, r = 30 μm and mass of 1.1E-10 kg is considered below. This requires that the millimeter size capsules will be designed to disintegrate into smaller capsules once within the target protostellar or accretion regions. For example, the 1 mm capsule may be made as a looser aggregate that will disintegrate by collisions with dust particles, or by evaporation of the binding matrix in the relatively warmer target zone, into of 30μm capsules. This particle size is comparable to the <1E-10 kg particles that constitute about 10% of the zodiacal cloud. Significantly, this particle size will not be ejected from the solar system by radiation pressure [14].

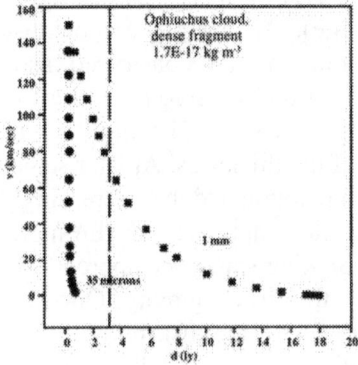

Fig. 2. The deceleration of 35μm and 1 mm radius objects inserted at a velocity of 1.5E5 m s^{-1} into representative regions of the Ophiuchus cloud. The objects are considered stopped at v = 2000 m s^{-1}. Note the decrease in penetration depth with density, and the relation to size of the cloud region (radii shows by vertical lines, for protostellar condensation r = 0.03 ly).

4. Targeting Strategies and Probability of Success

The fraction of launched panspermia swarm that will reach the target zone (the interstellar cloud, protostellar condensation etc.,) P_{target}, was calculated above. We consider here the further term $P_{capture}$, the probability that once in the target zone, the payload be eventually captured into the habitable zone of a planet. The overall probability for capture in the target planet is then obtained from equation (5).

$$P_{planet} = P_{target} \times P_{capture} \qquad (5)$$

As noted above, for calculated values of $P_{target} > 1$, we use $P_{target} = 1$. The following sections summarize the considerations to estimate $P_{capture}$, and from it, P_{planet}. The results are summarized in Table 1. The following discussion applies to solar sail missions (v = 5E-4 c), but results for advanced missions (v = 0.01 c) are also shown in Table 1.

4.1 Targeting the Dark Cloud Fragment

Equation (2) yields $P_{target} > 1$ for the dense Rho Ophiuchus cloud fragment. In other words, because of the large size of the target cloud, virtually all of the microbial capsules launched at it will arrive to the 3E16 m radius, 1E33 kg target. The cloud contains four dense cores with a total mass of about 1E31 kg, one of which has already formed protostellar condensations, and the others with the potential to form such condensations [8]. In addition, capsules may be also captured into the already formed 78 young stellar objects, which would have 100 au (1E13

286

m) radius dust shells or disks. Assuming that the cloud will eventually form 100 stars of 1E30 kg, from the mass ratio of each star to the overall dense cloud fragment, 1E-3 of the launched mass will be captured into each accreting solar system, i.e., for each star, P(target) = 1E-3. By the mass ratios of 1E17 kg dust captured by a planet during the suitable 1E9 yr prebiotic period to 2E30 kg mass of the protostellar condensation, about 1E-13 of the capsules will be captured, giving $P_{capture}$ = 1E-13. Altogether, therefore, P_{planet} = 1E-16 for each accreting solar system, i.e., 1E-16 of the mass launched at the cloud will be captured by a terrestrial planet in each accreting system. In total, 1E-14 of the launched mass will be captured in terrestrial type planets in the 100 accreting stars in this cloud. Note that with this strategy, individual stars are not targeted, and the mass that is launched must provide for seeding the entire cloud.

4.2 Targeting Individual Protostellar Condensations

Targeting individual protostellar condensations. The calculations above yielded P_{target} > 1 also for specific protostellar condensations, and therefore such regions can be targeted individually and we can use P_{target} = 1. From the mass balance ratios as above, $P_{capture}$ = 1E-13, giving also P_{planet} = 1E-13. The advantage of targeting individual protostellar condensations, rather than the overall cloud, is the greater chance for reaching a known, already established star-forming zone. This strategy also decreases the exposure time and radiation dose received when the payload would be diffusing through the cloud. A disadvantage is that, although the calculations yielded P_{target} > 1 for both the cloud and the individual protostellar condensations within it, the value was 1.4E4 for the cloud and only 1.4 for the condensation region, and realistically, the chances of capture are much greater in the larger cloud. Another disadvantage of targeting existing protostellar condensations is that the missions will miss many new star-forming condensations that form after the launching of the capsule swarm.

4.3 Targeting Early Accretion Disks

The 78 young stellar objects observed in Rho Ophiuchus are dust embedded or are in the T Tauri stage, with 100 au radius accretion disks. Because of their small size, P_{target} = 3.9E-3 for these objects. On the other hand, the capsules will be distributed only in the circumstellar dust but not in the star mass, avoiding a major source of loss. Assuming that the majority of the dust is accreted into the original 1E13 comets with a total mass of 1E28 kg, of which 1E17 kg is eventually captured by a planet, gives $P_{capture}$ = 1E-11, and P_{planet} = 3.9E-14.

4.4 Targeting Late Accretion Disks

Targeting accretion disks at the late stages of comet formation is advantageous because the capsules will be accreted into the outer cometary shell, which is most readily released subsequently. The theory of cometary accretion is uncertain, and a zone of some tens of au, say 10 - 20 au about the star may be considered for initial comet formation. For this area we obtain P_{target} = 1.2E-4. It will be assumed that the entire payload reaching the zone will be captured into orbit and eventually accreted into cometary shells. Assuming capture into the 100 m outer shell in 1E13 initial comets of 5,000 m radius, the microbial payload will be embedded in 3.1E26 kg dust, of which 1E17 kg will be delivered eventually to the planet, yielding P(capture) = 3.2E-10, and P(planet) = 3.8E-14.

4.5 Targeting Planets

The most direct approach is to target planets in already accreted planetary systems. As noted above, this may be better applied to planets at least 0.5 Gyr after accretion, as the initial conditions may be sterilizing. Targeting planets directly may be appropriate if older accreted planets are identified, or if further research suggests that young planets are survivable. We consider capture of the payload within <3.5 au from the star, which yields P_{target} = 4.9E-6. From the Zodiacal dust and meteorite capture statistics, $P_{capture}$ =1E-5, and therefore P_{planet} = 4.9E-11.

4.6 Biomass Requirements

The amount of material that needs to be launched is calculated from the P_{planet} values, allowing for the delivery of 100 capsules. The factor of 100 also corrects for other uncertainties in the mission. The mass required for the delivery of 100 capsules of 1.1E-10 kg each is then given by m = 1.1E-8/P_{planet}. The results are shown in Table 1.

For targeting the entire dense star-forming region, a very massive program of 1E8 kg per accreting star in the cloud is required, which can be only accomplished using space resources. If targeted at individual protostellar condensations or accretion shells or disks, requirements on the order of 1E5 kg for a sail mission, and especially 1E3 kg for an advanced mission, are realizable. Finally, if already accreted planetary systems in the cloud or closer are identified and targeted, the mass requirements on the <1 kg to 100 kg scale are easily met. Such panspermia programs should be affordable to small motivated groups or even individuals, which increases that likelihood that the program will be actually enacted.

4.7 Swarm Missions to Nearby Stars

Swarm missions to nearby stars. It is of interest to evaluate the swarm method also for closer planetary systems. For alpha PsA (Fomalhout), d = 22.6 ly, P_{target} was found as 1.2, and for beta Pictoris, 0.25, for capture into orbit in the habitable zone. For $P_{capture}$ we use 1E-5, although of course it may be different in different solar systems. With this assumption, P_{planet} = 1E-5 and 2.5E-6, respectively, is obtained for the two targets. These stars are in the local low-density interstellar medium, and the sail method described in the previous papers [4 - 6] may be used, miniaturized for launching 30 T m radius, 1E-10 kg capsules by small, 1.8 mm radius sails. These sails may be, for example, envelopes of thin reflective film that enclose the payload, mass-produced using industrial microencapsulation technologies. As few as 1E7 or 5E7 capsules, i.e., 1 or 5 g of microbial payload launched toward these stars in a swarm, respectively, could then deliver 100 capsules to a planet. Remarkably, with current launch costs of $10,000/kg, a panspermia swarm with a reasonable probability of success can then be launched to these stars, nominally, at the cost of $10. Of course, it should be easy to scale up such missions by a factor of 1,000 to kilogram quantities for increasing the probability of success or for allowing much less accurate, easier methods to launch the capsules, still within a very low-cost program of $10,000. Therefore, directed panspermia swarms to nearby planetary systems can be easy and inexpensive.

5. Survival and Growth in Comets and Asteroids

The missions to star-forming regions can arrive into solar systems at stars in various stages of star formation that may coexist in a target cloud. Stars that are at the dust-embedded or T Tauri stages when the missions are launched will last in these stages 1E5 - 1E6 years, similar to the transit time. When the missions arrive, these stars will have formed accretion rings. The subsequent planetary accretion lasts for 1E8 years, and high temperatures, intense solar UV flux, and frequent major impacts may make the new planets habitable only after 5E8 yr. However, capsules arriving at this stage can be preserved frozen if captured in asteroids and comets at r > 2.3 au at temperatures of T < 150 K, as calculated from the temperature function $T = 250r^{-0.6}$ (r distance in au). Furthermore, capsules accreted into a depth of several hundred g cm^{-2} in the comet will receive a radiation dose reduced by a factor of 100 from those on the cometary surface, which can assure survival on the Gyr time-scale.

Optimally, a fraction of the capsules may be embedded into the protected layers of the outer cometary crusts. These loose porous icy aggregates and embedded dust evaporate losing several hundred gm cm^{-2} in the first perihelion passage [11], and further inner layers evaporate gradually during further transits, releasing dust that is later captured into planets from the zodiacal cloud. Capsules that are more deeply embedded in cometary nuclei or asteroids may also arrive on planets with impacts [21], and within the meteorite rock can survive atmospheric transit.

Of the original 1E13 comets formed, 99% are ejected to interstellar space [12], but where Jupiter-sized planets fail to form, the cometary populations that remain bound to the solar system are greater, and barriers to penetration to crossing Earth-like planetary orbits are smaller. Jupiter-family comets can then remain in these orbits for 1E7 - 1E8 yr, instead of the present 1E5 yr, and the frequency of major cometary impacts increases from 1E-8 yr^{-1} to 1E-5 yr^{-1} [22]. In such planetary systems, the amount of cometary material and embedded microbial capsules that is delivered to the planets can increase by a factor of 1,000. In addition to comets, microorganism capsules may also become embedded in asteroids, and in the meteorites fragmented from them. Compared with the 1E26 kg total cometary mass, the total asteroid mass of 1E21 - 1E22 kg is much smaller, but it can provide a favorable nutrient microenvironment, see below.

6. Some Biological Considerations

The biological requirements were considered in relation to missions to nearby solar systems [4,5]. Some key points are as follows.

The microbial design must allow survival during transit, and subsequently in diverse planetary and possibly cometary environments, and facilitate evolutionary pressures that will lead to higher evolution.

These criteria suggest a diverse microbial assembly. The anaerobic environment will require at least facultative anaerobes. Blue-green algae, and possibly eukaryotic algae may be the best colonizing organism, the latter may lead to higher plant evolution. The photosynthetic organisms may survive first and establish an oxygen-containing atmosphere. Higher aerobes, including predatory heterotrophs can grow from the capsules that are meanwhile stored in comets and asteroids, and are delivered to the planet later. The ensuing predator/pray selection pressures will lead to higher evolution. This may require aerobic conditions, although conceivably, higher, including intelligent anaerobes may be possible.

The inclusion of simple multicellular eukaryotes is crucial, as this development may be a major evolutionary bottleneck. This development required billions of years on Earth, but then led rapidly to higher life-forms. Such a low probability event may not occur at all in other evolving ecosystems.

Even the most primitive single-cell organism must include the complex DNA and protein structures for replication, as well as complex energy mechanisms and membrane transport systems. The origin of such a complex system would seem to have a low probability. Panspermia helps to overcome this probability barrier. However, possible finding of Martian micro-organisms [26] may suggest that the origin of primitive life is more probable. Even in this case, overcoming the second probability barrier to the emergence of multicellular eukaryotes may in itself justify the panspermia program.

For interstellar transit, the microbial payload may be freeze-dried, as is the current practice for preserving microbial cultures. For UV survival, the capsules must be shielded appropriately, at least with UV resistant films. It may be also desirable to include a nutrient medium in the capsule, and to enclose it in a selective membrane that will allow the supplied nutrient to slowly absorb and mix with the local planetary nutrients, so that the microorganisms can gradually adjust to the planetary chemistry (pH, redox potential, toxic components, specific local nutrients). For aerobic eukaryotes, it may be desirable to enclose them in separate capsules with shells that will dissolve only in oxygen-containing environments. This will preserve the aerobic eukaryotes until photosynthetic organisms create a suitable oxygen-containing atmosphere.

It may be possible to provide some of this shielding and nutrient using the solar sail that launches the capsule. The sail must constitute about 90% of the total mass of the small vehicles. The sail could be possibly made of proteinaceous or other biodegradable organics. It may be designed to fold over the microbial packets after propelling them from the solar system, and provide shielding during transit and capture, and eventually to provide nutrient materials on the host planet.

For successful missions, the microorganisms must find adequate nutrients, which may be carbonaceous materials accumulated from dust particles, comets and asteroids, with organic content resembling carbonaceous chondrites. As a model, soil nutrient analysis of the Murchison C2 meteorite showed biologically available nutrient content (in mg/g) of: C and N in hydrothermally (121 °C, 15 minutes) extractable organics, 1.8 and 0.1; S as soluble SO_4^{-2}, 4.5; P as PO_4^{-3}, 6.4E-3; and extractable cations by 1 M CH3COONH4 solution at pH 7, Ca, 4.0; Mg, 1.7; Na, 0.57; K, 0.65 mg/g; and cation exchange capacity of 5.8 milliequivalents/100 g. All of these are values are comparable or higher than in average terrestrial agricultural soil. Use of the organic meteorite nutrients as sole carbon source was demonstrated by light emission from *Pseudomonas fluorescence* modified with a lux gene when challenged with the meteorite extract, and preliminary observations of growth of the thermophile eubacteria *Thermus* and *Thermotoga* in the extract. The soil microorganisms *Flavobacterium oryzihabitans* and *Nocardia asteroides* grew in materials extracted from 100 mg meteorite powder into 1 ml water to populations up to 5E7 colony forming units/ml in 4-8 days, similar to extracts from agricultural soils, and retained stable populations in the meteorite extract for several months. Biological effect on higher plants was demonstrated by *Asparagus officinalis* and *Solanum tuberosis* (potato) tissue cultures. When the above meteorite extract was added to partial 10 mM $NH_4H_2PO_4$ nutrient solution, the average fresh weight of asparagus plants grew from 1.5±0.3 to 2.1±0.8 g, and of potato from 3.0±1.2 to 3.9±1.2 g, and both showed enhanced green coloration. Correspondingly, the elemental S content of asparagus dry mass increased from 0.07 to 0.49%, of Ca from 0.02 to 0.26, of Mg from 0.03 to 0.41, of K from 0.18 to 0.32, of Fe from 0.02 to 0.03% [18,19].

These observations suggest that microorganisms entering young planetary environments, and even higher organisms, can grow on the large amounts of accreted interplanetary dust, meteorite and cometary [23] materials. Implanted microorganisms may multiply as well in carbonaceous asteroid parent bodies during the warm hydrothermal alteration phase, and in dust-sealed comets if they contain sub-surface water when warmed to 280-380 K during perihelion transits [27]. After landing, microorganisms can use the meteorite matrix materials. In fact, water in fissures in carbonaceous meteorites can create concentrated organic and mineral nutrient solutions conducive to prebiotic synthesis, and provide early nutrients after life arose in these meteorite microenvironments [19].

7. Advanced Missions and Development Needs

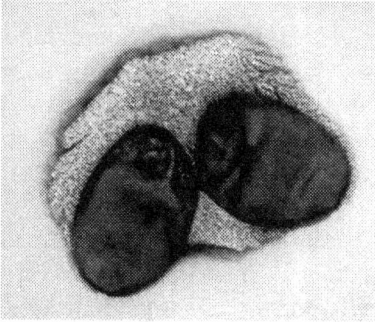

Fig. 3 Microorganisms identified tentatively as *Flavobacterium oryzihabitans* grown in extraterrestrial nutrient extracted from the Murchison meteorite, with a meteorite fragment in the background. 5),

Reprinted from M. N. Mautner, R. R. Leonard and D.W. Deamer, "Meteorite Organics in Planetary Environments: Hyrothermal Extraction, Surface Activity and Microbial Utilization", Planetary and Space Science 43, 139-147 (199

Advanced technologies can increase substantially the probability of success, and reduce the required swarm mass, which is a major economical barrier.

(1) Preparation of Biological Payload. Genetically engineer microorganisms, including multicellular eukaryotes that combine extremophile traits for survival in unpredictable, diverse environments and that can efficiently metabolize extraterrestrial nutrients. It may be necessary to devise missions where the microbial payload can defrost and multiply/recycle periodically, say every 1E5 yr, for renewal against radiation-induced genetic degradation.

(2) Propulsion. Develop new methods to accelerate sub-milligram objects to 0.01 c. For example, antimatter - matter recombination has the potential to reach velocities close to c. Interestingly, the energy for a capsule of 1E-6 kg travelling at 0.01 c, i.e., 4.5E6 J, which can be provided by mass-to-energy conversion of 5E-11 kg of antiparticles. Launching smaller, microgram capsules at 0.01 c requires the production of 5E-14 kg of antiparticles, which brings even this exotic technology within the capabilities of current technology [24].

(3) Navigation. Apply on-board intelligent robots for in-course navigation, and for identifying suitable accretion systems and

habitable planets; for landing on these targets; and to control the initial incubation.

(4) Accretion into comets and asteroids. Use self-replicating robots to multiply on those bodies and to turn them into biological hatcheries. Use comets and asteroids in this solar system to grow large panspermia biomasses for interstellar and galactic panspermia, and as growth and storage media in the target systems.

At the highest technological level, human interstellar travel can promote life. For example, Oort-belt cometary nuclei can be converted to habitats with resources to sustain each up to 1E13 kg biomass (1E12 human population), and their large-aphelion orbit readily perturbed to leave the solar system. Human interstellar travel may require centuries of far-reaching developments, including the bioengineering of space-adapted, science-based "homo spasciense". Space adaptation may also need man/machine cyborgs and the risk of robot takeover, or strong measures to ensure that control stays vested in organic intelligent brains with self-interest in perpetuating their (and our) genetic heritage as DNA/protein life.

Such problems illustrate that human interstellar travel is tenuous. The longevity of intelligent civilizations is unknown, and the long-term ability of organic intelligent Life to propagate itself in space is unpredictable. It is therefore prudent to enact a panspermia program early using available technology, and advanced technologies can be incorporated as they develop.

8. Resource Requirements For Seeding the Galaxy

Although aimed at specific targets, the microbial payloads may carry life further in space and time.

First, much of the microbial swarm will miss or transit the target. Secondly, of the initial 1E13 comets that capture capsules in the accreting system, up to 99% will be ejected into interstellar space [11], carrying the microbial content. These embedded capsules, shielded from radiation and preserved at 3 K, may survive in the comets for many Gyr, until eventually captured in accreting systems in other regions of the galaxy. Of the perhaps 1E11 comets remaining in the accreting system, most will remain in the cold <10 K Oort cloud which will be eventually ejected into interstellar space. Therefore the majority of the launched biomass will eventually carry the microbial payload further into the galaxy. The spread of microbial life by comets is similar to the proposals of Hoyle and Wickramasinghe [17], but we postulate here a directed origin.

Future programs may aim intentionally to seed the entire galaxy. It is interesting to assess the feasibility of such a program.

Once launched randomly into the galactic plane at v = 0.01 c, the microbial packets will traverse the galaxy (r = 7E4 ly [25]) in 7E6 yr. The packets are gravitationally bound to the galaxy and will eventually perform random paths. At these speeds, mm size capsules will transit all thin regions and will be captured only in protostellar condensations or denser accretion zones. The mass ratios above showed that 1E-13 of the captured biomass in these areas will be delivered to planets. With 100 capsules of 1E-10 kg, i.e., a biomass of 1E-8 kg required to seed a planet, and with star-formation rate of 1 yr^{-1} in the galaxy, biomass needs to be launched at the rate of 1E5 kg/yr for 5E9 yr to seed all new stars during the lifetime of the solar system. For example, the biomass can be dispersed in pulses of 1E12 kg to seed the population of star-forming clouds as it is renewed every 1E7 yr. The total required biomass is 5E14 kg, compared for example with the 1E19 kg organic carbon (1%) in the 1E21 kg total asteroid mass. This resource allows increasing the launched biomass up to a factor of 2E6 to account for undoubtedly substantial losses.

As a more conservative estimate, assume a 5 au capture zone, with a volume of 2E36 m^3, with the total capture volume of 2E47 m^3 about 1E11 stars. With a capture probability of 1E-5 and for delivering 100 captured capsules of 1E-10 kg each, 1E-3 kg needs to be placed about each star. This corresponds to a density of 5E-40 kg biomass m^{-3} in these circumstellar volumes. Assuming that this is achieved by establishing a similar biomass density through the 5E61 m^3 volume of the galaxy, then the total biomass needed in the galaxy is 2.5E22 kg. Renewing this density each 1E9 yr for the 5E9 yr lifetime of the solar system, to seed every new planetary system during the first Gyr after its formation, gives a material requirement of about 1E23 kg, about 10% of the 1% C content in 1E26 kg of the total cometary mass.

The material requirements can be reduced by many orders of magnitude if the missions are directed to star-forming regions rather than distributing biomass through the galaxy at random. Of course, the microbial population may be subject to substantial losses, but may be enhanced in the target zones by gravitational attraction. The fate of biological objects traversing the galaxy requires detailed analysis.

It may be possible to grow the necessary large amounts of microorganisms directly in carbonaceous asteroids or comets.

Carbonaceous C1 meteorites, and presumably asteroids, contain water in about the biological ratio of 5:1 H_2O/C, and N in the biological ratio of 10:1 C/N, as well as biologically usable forms of the other macronutrients S, P, Ca, Mg, Na and K in at least the biological C/X elemental ratios [19]. Once the nutrient components are extracted, the residual inorganic components may be used for shielding materials for the microbial capsules.

As a possible method for converting comets to biomass, the loose icy, cometary matrix may be fragmented and enclosed in membranes in 1 kg spheres. Warming and melting such a unit, from 10 to 300 K, requires 5.1E9 J, which can be provided by the solar energy flux of 325 W m^{-2} at 2 au, incident on the 3.1 m^2 cross-section of a 1 m radius object during a two-months perihelion transit about 2 au. The microbial experiments show that in 6 - 8 days after inoculation, this organic solution will yield microbial densities of >1E8 CFU/ml which can survive for several months [18, 19]. Subsequently, the microbial solution can be converted to 1 mm "hailstones". These microbial ice capsules can be accelerated out of the solar system, for example, by first accelerating the comets sunward into parabolic orbits, and in this manner dispersing the Oort cloud at the rate of 20 comets yr^{-1} during 5E9 yr. This rate is comparable to the natural rate of 3 new comets/yr plus up to 1E9 new comets per/year during cometary showers [16], and the task may be accomplished at the required rate by processing every new comet that arrives naturally from the Oort cloud.

An interesting experiment in this direction would be to inoculate the sub-crust zone of an inbound comet, and of enclosed samples of the cometary material embedded in the comet, the latter to allow melting near the perihelion without evaporation. Embedded sensors could monitor microbial growth during the perihelion passage, and in short-period comets during further passages, to verify microbial growth in cometary materials and environments. Laboratory microbiology experiments with returned cometary materials would be also of interest.

The above considerations suggest that a single technological civilization can seed the galaxy. Similarly, one past panbiotic civilization could have seeded the galaxy, accounting for the rapid emergence of life on Earth and possibly on Mars [2, 3, 26].

By extrapolation, the material resources of 1E11 solar systems in one galaxy may be sufficient to seed all the 1E11 galaxies.

9. Motivation: The Principles of Panbiotic Ethics

Directed panspermia must rest entirely on enduring ethical motivation. Eventually, this non-material moral entity can have far-reaching consequences on the material future of the universe.

The insights of contemporary biology and cosmology can be synthesized into a Life-centered panbiotic ethics, as follows.

(1) Life is a process of the active self-propagation of organized molecular patterns.

(2) The patterns of organic terrestrial Life are embodied in biomolecular structures that actively reproduce through cycles of genetic code and protein action.

(3) But action that leads to a selected outcome is functionally equivalent to the pursuit of a purpose.

(4) Where there is Life there is therefore a purpose. The object inherent in Life in self-propagation.

(5) Humans share the self-propagating DNA/protein biophysics of all cellular organisms, and therefore share with the family of organic Life a common purpose.

(6) Assuming free will, the human purpose must be self-defined. From our identity with Life derives the human purpose to forever safeguard and propagate Life. In this pursuit human action will establish Life as a governing force in nature.

(7) The human purpose defines the axioms of ethics. Moral good is that which promotes Life, and evil is that which destroys Life.

(8) Life, in the complexity of its structures and processes, is unique amongst the hierarchy of structures in Nature. This unites the family of Life and raises it above the inanimate universe.

(9) Biology is possibly only by a precise coincidence of the laws of physics. Thereby the physical universe itself also comes to a special point in the living process.

(10) New life-forms who are most fit survive and reproduce best. This tautology, judgement of fitness to survive by survival itself, is the logic of Life. The mechanisms of Life may forever change, but the logic of Life is forever permanent.

(11) Survival is best secured by expansion in space, and biological progress is best assured by adaptation to diverse multiple worlds. This process will foster biological and human/machine coevolution. In the latter, cont5rol must always remain with organic-based intelligences who have vested interests to continue our organic life-

form. When the future is subject to conscious control, the conscious will to continue Life must itself be forever propagated.

(12) The human purpose and the destiny of Life are intertwined. The results can light up the galaxy with life, and affect the future patterns of the universe. When the living pattern pervades nature, human existence will have attained a cosmic purpose.

Points 3-5 do not suggest teleology, i.e., it is not implied that the biological process recognizes an objective. Rather, these points are based on the principles of equivalence, that also underlie, for example, relativity and Turing's test of intelligence: if an entity is indistinguishable in all observables from another entity, then the two are identical. Applied here, if the biological process was seeking to propagate purposefully, it would function as it does actually. Therefore the behavior of the biological process is indistinguishable from, and equivalent to, action with purpose.

A serious panbiotic motivation in our young civilization is expressed by the society for the Interstellar Propagation of Life [28]. Depending on the frequency of life in space, the panspermia program may have the following objectives.

(1) The complex mechanism of replication, transcription, energy production and membrane transport must be all present in even the simplest surviving cell. This crates a large probability barrier to the origin of life. If terrestrial life is alone, we have a special duty to safeguard and propagate this unique creation of nature.

(2) Eukaryotic and multicellular life emerged on Earth only after 3E9 yr. This shows a large probability barrier to higher evolution, which therefore may not occur at all in other primitive biosystems. Eukaryotes and simple multicellular organisms that could survive interstellar transport should be included in the panspermia payload to overcome this evolutionary barrier. A possible outcome of the resulting higher evolution, as our own panbiotic capabilities demonstrate, is the emergence of new intelligent species who will promote Life further in the galaxy.

(3) Extraterrestrial intelligent life is counter-indicated by a lack of scientific evidence and by Fermi's paradox. Should such civilizations exist, however, panspermia can serve as interstellar communication. The nature of our life form, that we cannot yet fully describe, is best communicated by samples. If our life form will have to compete with others, we shall have only extended this basic property of the living process. Our innate duty is first to our own organic life form.

The technical approaches to panspermia will evolve, but a permanent ethical foundation must prevail. The biocentric principles derive rationally from the scientific world-view, and are also consistent with the respect for life innate in healthy human emotions, civilizations and religions. The Life-centered panbiotic purpose to propagate Life can therefore serve as a lasting basis of human ethics.

The panbiotic enterprise will transform new solar systems through the galaxy into evolving biospheres. In this process, Life will achieve secure continuation, the diversification of species and even higher patterns of complexity. Eventually, all the material constituents of nature, as it extends in time and space, will become living substance and its sustaining resources. Once we plant life in space, the self-propagating nature of Life will assure that it will encompass all matter. In this sense, the physical universe itself will have become an interconnected living being.

When Life comes to the universe, the universe will come to life. In fulfilling the ultimate purpose of Life, our human existence will have assumed a cosmic meaning.

References

1. S. Arrhenius, "Vernaldas Ultveckling", Stockholm, 1908. Quoted by A. I. Oparin in "Genesis and Evolutionary Development of Life", Academic Press, New York.
2. Shklovskii and C. Sagan, "Intelligent Life in the Universe", Dell, 1966.
3. F. H. Crick and L. E. Orgel, "Directed Panspermia", Icarus 19, 341, 1973.
4. M. Meot-Ner (Mautner) and G. L. Matloff, "Directed Panspermia: A Technical and Ethical Evaluation of Seeding Nearby Solar Systems. J. British Interplanet. Soc. 32, 419-423 (1979).
5. M. N. Mautner, "Directed Panspermia. 2. Technological Advances Toward Seeding Other Solar Systems, and the Foundations of Panbiotic Ethics", J. British Interplanet. Soc. 48, 435-440 (1995).
6. M. N. Mautner and G. L. Matloff, "An Evaluation of Interstellar Propulsion Methods for Seeding New Solar Systems", Proceedings of the First IAA Conference on Realistic Deep-Space Missions, Turin, Italy, June 1996, Levrotto and Bella, Turin (1996).
7. B. Zuckerman, "Space Telescopes, Interstellar Probes and Directed Panspermia", J. British Interplanet. Soc. 34, 367-370 (1979).

8. P. G. Mezger, "The Search for Protostars Using Millimeter/Submillimeter Dust Emission as a Tracer", in Planetary Systems: Formation, Evolution and Detection, ed. B. F. Burke, J. H. Rahe and E. E. Roettger, Kluwer Academic Publishers, Dordrecht, p. 197-214, 1994.

9. V. S. Safronov and E. L. Ruskol, "Formation and Evolution of Planets", in Planetary Systems: Formation, Evolution and Detection, ed. B. F. Burke, J. H. Rahe and E. E. Roettger, Kluwer Academic Publishers, Dordrecht, p. 13-22, 1994.

10. R. Neuhauser and J. V. Feitzinger, "Radial Migration of Planetisimals", in Planetary Systems: Formation, Evolution and Detection, ed. B. F. Burke, J. H. Rahe and E. E. Roettger, Kluwer Academic Publishers, Dordrecht, p. 49-56, 1994.

11. M. E. Bailey, S. V. M Clube and W. M. Napier, "The Origin of Comets", Pergamon Press, Oxford, 1990, p. 435.

12. J. H. Oort, "Orbital Distribution of Comets", in W. F. Huebner, ed. "Physics and Chemistry of Comets", Springer-Verlag, Berlin, 1990.

13. J. Davies, "Frozen in Time", New Scientist p. 36-39, 13 April, 1996.

14. D. Morrison, "Sizes and Albedos of the Larger Asteroids", in Comets, Asteroids and Meteorites, Interrelations, Evolution and Origins, A.H. Delsemme, ed., U. of Toledo Press, p. 177-183, 1977.

15. Z. Sekanina, "Meteor Streams in the Making", in Comets, Asteroids and Meteorites, Interrelations, Origins and Evolution, A.H. Delsemme, ed., U. of Toledo Press, p. 159-169, 1977.

16. A. F. T. Kyte and J. T. Wasson, "Accretion Rate of Extraterrestrial Matter: Iridium Deposited 33 to 67 Million Years Ago", Science 232, 1225-1229 (1989). b. G. W. Wetherill, "Fragmentation of Asteroids and Delivery of Fragments to Earth", in Comets, Asteroids and Meteorites, Interrelations, Evolution and Origins, A.H. Delsemme, ed., U. of Toledo Press, p. 283-291, 1977.

17. F. Hoyle and C. Wickramasinghe, "Lifecloud: the Origin of Life in the Universe", J. M. Dent and Sons, London, 1978.

18. M. N. Mautner, R. L. Leonard and D. W. Deamer, "Meteorite Organics in Planetary Environments: Hydrothermal Release, Surface Activity and Microbial Utilisation", Planet. Space Sci., 43, 139-147 (1995)

19. M. N. Mautner, "Biological Properties of Meteorite Materials: Macronutrients and Growth in Carbonaceous Chondrites", submitted, 1996. Observations for Pseudomonas fluorescence from K. Killham and for Thermus and Thermotoga from H. W. Morgan, for plant cultures from A. J. Conner, private communication, 1996.

20. E. Anders, "Prebiotic Organic Matter from Comets and Asteroids", Nature 342, 255-257 (1989).

21. T. Owen, "The Search for Other Planets: Clues from the Solar System", in Planetary Systems: Formation, Evolution and Detection, ed. B. F. Burke, J. H. Rahe and E. E. Roettger, Kluwer Academic Publishers, Dordrecht, p. 1-11, 1994.

22. G. W. Wetherill, "Possible Consequences of Absence of "Jupiters" in Planetary Systems, in Planetary Systems: Formation, Evolution and Detection, ed. B. F. Burke, J. H. Rahe and E. E. Roettger, Kluwer Academic Publishers, Dordrecht, p. 23-32, 1994.

23. B. C. Clark, "Primeval Procreative Comet Pond", Origins of Life 18, 209-238 (1988).

24. J. H. Mauldin, "Prospects for Interstellar Travel", AAS Publications, Univelt, San Diego, 1992.

25. M. Zeilik, S.A. Gregory and E. P. Smith, "Astronomy and Astrophysics", Saunders, Fort Worth, p. 383, 1992.

26. 2D. S. McKay, E. K. Gibson, K. L. Thomas-Kerpta, H. Vali, C. S. Romanek, S. J. Clemett, X. D. F. Chillier, C. R. Maechling and R. N. Zare, " Search for Past Life on Mars: Possible Relic Biogenic Activity in Martian Meteorite ALH84001", Science 273, 924-930 (1996)

27. Chyba, C. F. and McDonald, G. D. "The Origin of Life in the Solar System: Current Issues", Annu. Rev. Earth Planet. Sci., 1995, 23, 215-249.

28. M. N. Mautner. "Society for the Interstellar Propagation of Life (Interstellar Panspermia Association)", founded 1996. Announcement in the electronic exobiology newsletter "Marsbugs", J. Hiscox and D. Thomas, eds, 1995. Internet contact of SIPL through mautnerm@lincoln.ac.nz.

Appendix A3.6

AAS 89-668

SPACE-BASED CONTROL OF THE CLIMATE

Engineering, Construction, and Operations in Space II, Proceedings of Space 90, S. W. Johnson and J. P. Wetzel, Editors, Albuquerque, New Mexico, April 22-26 1990, pp. 1159-1168.

Michael Mautner
[1]Department of Chemistry, University of Canterbury, Christchurch 1, New Zealand; [2]American Rocket Co. Camarillo, California 93010.

Abstract

The expected global greenhouse warming of 2° C may be reversed by a space-based screen that intercepts 3% of the incident solar radiation. Space-based devices may also mitigate the ozone/UV problem and extreme weather phenomena such as cyclones. Against greenhouse warming, a fleet of screen units can be deployed in Earth orbit, at the sun-Earth L1 point, or in radiation-levitated solar orbit. For the latter, the required areal density at 0.1 au from the Earth is 5.5 g m^{-2}, and the total required mass is 1.7×10^{10} kg, equivalent to a hill or asteroid with a radius of 120 m. The required material can be obtained from lunar sources by mass drivers. Active orbital control can be achieved using solar sailing methods. In the long run, similar devices may also modify the climates of other planets for terraforming. Space-based climate control is possible through the use of large energy leverage, i.e., solar screens and reflectors manipulate 10^6 times the energy required for deployment. Therefore in the space scenario projected for several decades, substantial intervention with the climate will be facile, and this suggests serious studies.

Appendix A3.7

AAS 89-668

DEEP-SPACE SOLAR SCREENS AGAINST CLIMATIC WARMING: TECHNICAL AND RESEARCH REQUIREMENTS

Michael Mautner
Department of Chemistry, University of Canterbury, Christchurch 1, New Zealand

Space Utilization and Applications in the Pacific
Proceedings of the Third Pacific Basin International Symposium on Advances in Space Sciences Technology and its Applications (PISSTA) held November 6-8, 1989, Los Angeles, California. American Astronautical Society Advances in the Astronautical Sciences volume 73, PM Bainum, GL May, T Yamanaka, Y Jiachi, editors, Univelt, Inc. San Diego, CA 1990

Intercepting about 3% of the Earth-bound solar radiation in space can prevent the expected greenhouse warming and surface UV incidence. Solar screen units based on thin-film material can be deployed in levitated solar orbit or about the Sun-Earth L1 libration point. The mass requirements are 10^9 - 10^{11} kg which can be obtained from lunar or asteroid sources and processed using vacuum vapor deposition. The screen could prevent the projected \$10 trillion economic damage of climatic warming by 1% of the cost. Using screens of dust grains could further reduce the deployment and operation costs by an order of magnitude. Cost/benefit analysis justifies a \$4B research program at present and a \$100B program for deployment 40 years prior to the onset of serious climatic warming. Most importantly, such screens should be studied as a measure of last resort if the Earth becomes threatened by runaway greenhouse warming due to positive feedback effects.

Introduction

It may become necessary to control the climate of the Earth or the planets in order to keep, or render, the environment habitable. For example, the climate of the Earth is expected to change because of increasing levels of CO_2 and other greenhouse gases in the atmosphere. The resulting greenhouse effect is likely to lead to climatic warming by 1 - 5 C by the middle of the next century, with flooding, droughts, wildlife extinction, and economic damage on a global scale [1 - 3]. These problems are compounded by increased UV radiation on the Earth's surface because of the depletion of the ozone layer. The Antarctic ozone hole is already increasing UV exposures in the South Pacific regions of Australia and New Zealand.

Environmental solutions to these problems are desirable but costly. For example, developing renewable sources for the world's energy needs may cost trillions of dollars [4-6]. If environmental solutions fail, the damage by the greenhouse effect would also amount to trillions of dollars [5].

We recently pointed out that a climatic crisis may be averted by decreasing the solar radiation incident on the Earth in the first place [7]. This may be achieved using existing or near-future space technology, with expenses much less than the cost of greenhouse damage. However, such intervention should be done cautiously since the ultimate effects of climatic intervention may be hard to predict [5].

As intervention with the climate is becoming feasible, it should be seriously studied and debated. Designing the space-based solar screen is an interdisciplinary project involving astronomy, astronautics, optics, material science, geophysics, atmospheric science, climatic modeling, ecology, economy and international law. This paper will present some general considerations in several of these areas, and survey specific research needs.

a. Climatic Effects

The relation between insolation and climate have been investigated by several authors [8-11]. In brief, the most pertinent results from general circulation models are that: (1) The heating effect by doubling the atmospheric CO_2 is $Q_{inc} = 4$ W m^{-2}, i.e., equivalent to increased insolation by 3%; (2) The resulting temperature rise is determined by $T_{inc} = Q_{inc}/L$, where L is the feedback factor depending on feedback effects caused by such effects as increased atmospheric humidity and decrease albedo by melting ice-covers. Considering the uncertainty in L give $T_{inc} = 1.6$ to 4.4 K for doubled atmospheric CO_2 [12].

More research is needed to predict the climatic effects of solar screens.

- More accurate models are needed for the relation between insolation and temperature. In fact, effects of varying the incident solar radiation by several per cents do not seem to have been modeled to date.

- Climate models are needed that allow simultaneously for increased CO_2 and decreased solar radiation. Regional effects also need to be modeled to evaluate the effect of climatic screen on various nations.
- The atmospheric effects of decreased solar UV need to be considered to predict the effects of the UV filter screens described below. In particular, atmospheric ozone itself is formed by solar UV induced radical reasons, and the UV screen could cause a cycle of further ozone depletion and further screening.

b. Screen Designs

A primary consideration for the feasibility of a space-based solar screen is the required mass. Considering the Sun as a point source at infinite distance, and a need to screen out 3% of the incident solar radiation, the area covered by a screen in the Earth's vicinity is $0.3 \times 3.14 \times R_{Earth}^2 = 3.8 \times 10^{12}$ m^2. Using a screen of thin film, supporting mesh and steering mechanisms with a compounded areal density of 1 g/m^2, the required mass is 3.8×10^9 kg, equivalent to a boulder with a radius of 68 m. As we shall see, the actual mass may change somewhat with screen design. In any event, these simple considerations show that a small mass properly deployed in space can modify the climate to an extent comparable with the expected greenhouse effect or even with a natural ice age [13].

For structural reasons, any screen of the required size will have to be made of a large number of independent units, all of which combined are referred to here as "the screen". One possible configuration would be a belt of screen units in Earth orbit. Using 100% radiation absorbing material and accounting for the radiation balance between the Earth and the screen leads to a minimum mass requirement at an altitude of 2,200 km, where a structure with an overall areal density of 1 g/m^2 required 3.4×10^{10} kg. [7]. In comparison, constructing a totally reflective film at a distance from the Earth where mutual re-absorption and re-radiation between the Earth and screen is negligible requires a mass of 1.6×10^{10} kg [7]. These masses may have to be increased by a large factor, or substantial propellant mass flows comparable to the L1 screen described below may need to be employed to balance against radiation pressure. Furthermore, a screen in Earth orbit may interfere with astronautics and astronomy, may suffer from residual atmospheric drag, and may lead to uneven thermal screening of the Earth's surface.

An alternative that can be deployed deeper in space is a screen of levitated solar sails [14-16]. In this case the solar radiation pressure allows the screen to be placed in a solar orbit synchronous with the Earth, but closer to the Sun, permanently on the Sun - Earth axis. The balance of forces on the sail shows that the required overall areal density is a function only of the Sun - screen distance. For example, at 0.1 au from the Earth, the required density turns out to be 5.5 g/m^2, a technologically convenient value. Details of levitated solar screens are summarized in Table 5.1.

This paper will deal in some detail with a third alternative, screens deployed about the Sun-Earth L1 point, which is 0.01 Au from the Earth. The screens must reflect or deflect 3% of the Earth-bound solar flux or to absorb the total Earth-bound solar UV.

The L1 space position was used for the International Sun-Earth Explorer-3 (ISEE-3) libration point satellite, which was deployed in a halo orbit [17,18]. This deployment did not depend on solar sailing, and the areal density was not a limiting factor.

To shade the Earth the screen must be on the cross-sectional area that occults the Earth. Considering the Sun as a point source at infinite distance, the required area has a radius of 6.4×10^6 m (equal to R_{Earth}) in the plane perpendicular to the ecliptic (the x, y plane).

The screen units may be kept in the required area by orbiting the L1 point. This is similar to the halo orbits of the ISEE-3 satellite. However, in that case the objective was to keep the satellite at least 6 degrees away from the Sun-Earth axis, and a large y amplitude (700,000 km) was used [17,18]. The present requirements are opposite, as the screen units must be within 0.5° of the Sun-Earth axis and in an orbit with the x and y amplitudes of 6,400 km or less. The stability of such orbits and the injection velocity costs for objects into such orbits need to be studied. If small orbits about the L1 point are not practical, then the screen sections may be kept stationary by thrusters.

c. UV Filter Screen

It may become desirable to screen out the solar ultraviolet (UV) incident on Earth in the range 250-340 nm. This would happen if the ozone loss becomes serious. This part of the solar spectrum constitutes about 4% of the incident solar radiation flux, and eliminating it would also help to decrease greenhouse warming (although this is not equivalent to a 4% screening of the entire spectrum since most of the UV does not penetrate to the Earth's surface).

UV levels are increasing by up to 5% in some areas. A seriously worsening situation may raise this to 10% globally. To counter this effect it would be necessary to pass 10% of the Earth-bound radiation through the filter that absorbs 100% of the incident UV. Therefore the area is $0,.1 \times 3.14 R_{Earth}^2 = 1.3 \times 10^{13}$ m^2, and with d = 1 g m^{-2}, the total mass of the screen is 1.3×10^{10} kg.

For this filter screen, the interaction factor is a = 0.04. Using mass drivers, the required $m_{fp} = 3.10^{10}$ kg/y and with ion thrusters, $m_{fp} = 1.5 \times 10^9$ kg/y. In this case $W_p = 5.8 \times 10^{19}$. Again, the filter screen cannot be used as a collector, and for a conversion efficiency of 0.1, the power requires a collector surface of 4.2×10^8 m^2. Here again the propellant mass flow is the limiting factor.

Table 5.1. Technical Requirements for Space-Based Screen for Climate Control

	Screen-ing Effect[a]	Total Area (m^2)	Total Mass (kg)	Areal Density (kg m^{-2})	m_{fp}[b] (kg/yr.$^{-1}$)	W_p[c] (W)
Reflective belt in Earth orbit at 2,200 km[d]	3% flux	1.6×10^{13}	1.6×10^{10}	10^{-3}	(a)	(a)
Absorbent belt in Earth orbit at 2,200 km	3% flux	3.4×10^{13}	3.4×10^{10}	10^{-3}	(a)	(a)
Levitated sail at 0.1 AU	3% flux	3.1×10^{12}	1.7×10^{10}	5.5×10^{-3}	(o)	(o)
Levitation-supported UV-screen	3% flux, 15.5% UV	$1.7 \times 10 + ^{13}$	9.4×10^{10}	5.5×10^{-3}	(o)	(o)
Reflective sail at L1	3% flux	3.8×10^{12}	3.8×10^9	10^{-3}	2.2×10^{10}	8.6×10^{11}
Deflection screen at L1	3% flux	3.8×10^{12}	3.8×10^9	3.8×10^9	9.5×10^4	3.8×10^6
UV screen at L1	10% flux	1.3×10^{13}	1.3×10^{10}	10^{-3}	1.5×10^9	5.810^{10}

Footnotes to Table 5.1.

a. "3% flux" means screening out of 3% of the total solar flux, 15.5% UV means screening out 15.5% of the solar UV flux, etc.
b. Rocket thruster propellant for momentum balance against solar radiation pressure with v_p = 5×10^4 m sec^{-1}. Mm$_{fp}$ is the total propellant mass flow required for momentum balance for the entire screen.
c. Total power for propellant drives. Propellants may be necessary to stabilize the orbits against radiation pressure effects.
d. Screen at 0.1 au from the Earth, constituted of 1 m^2 of 1 g m^{-2} reflective film levitating 4.5 m^2 of 1 g m^{-2} UV screen.

There are alternatives to using thin film technology for the solar screen. The screen could be constituted of fine wire mesh or even fine glass fibers that scatter 3% of the incident radiation, supported on a network of heavier wire and beams. The mesh would spread over the cross-sectional area of the Earth but its actual surface area would cover only 3%. The overall mass, radiation pressure and propellant requirements would be similar to the film screen, but manufacturing may be easier. If such a screen scatters rather than reflects the incident radiation, the radiation pressure and required balancing propellant flow could be substantially reduced.

d. Dust Grain Screens

A major part of the costs of thin film or optical filter screens arises from the manufacturing of the materials and structures, and subsequent maintenance. Most of these costs could be eliminated if the solar radiation could be screened out by a cloud of particulate dust. Such dust could be generated from finely ground but otherwise unprocessed lunar dust or asteroidal rock transported to space as loosely bound agglomerations and gently dispersed into the desired orbit with small lateral velocities. As for mass requirements, the total geometrical cross-section of micro-sized grains is comparable to a 1 g m^{-2} film with the same surface area [7]. Larger grains will require larger masses [7], but will be more stable in orbit. Using such grain sizes, most of the solar radiation will be scattered, possibly by small angles from the incident radiation. As noted above, deflection by 0.24° at L1 is enough to divert photons from the Earth.

Dust clouds could be placed at L1, or in a levitated orbit if the radiation pressure can be balanced against other forces to support such orbits. The primary question is the stability and dispersion rate of the cloud at either location. This is a complex phenomenon which is affected by collisions, radiation pressure, the Poynting-Robertson effect, the solar magnetic field, solar wind and gravitational effects. In general, the lifetime of particles in the solar system decreases with decreasing size. Article survival rates at 1 au from the Sun at of 10^3 - 10^4 years [24], but the important question is how long would the particles remain in the 6,4000 km radius area required to shade the Earth. It is readily shown that thermal velocities of micron-size particles, or motion due to the small forces at libration points or levitated orbits would not lead to acceptable collision rates. It is possible that such clouds could be charged and then controlled electrostatically.

If the required dust clouds have masses comparable to film screen, i.e., 10^9 - 10^{11} kg, then it is desirable that the dispersion rates will be at most 100% - 1% per year, respectively. In this case replacement could be made using one lunar mass driver, at a small cost.

The dust cloud approach can be combined with other technologies, if necessary. For example, a 1 km diameter lunar boulder, asteroid or equivalent mass could be stationed at L1 or in levitated orbit, and slowly ablated to maintain a tail of dust that occults the Earth. At the rate of 10^{10} kg per year, such a source would maintain a screen for 100 years.

e. Research Requirements for Dust Screens

The behavior of interstellar dust clouds is a complex but well-researched subject. Research is needed on the stability of dust clouds of the required mass and density; the optimal grain size; the possibility of such clouds in levitated orbits; techniques for electrostatic control. It would be particularly interesting if grains of the appropriate size could scatter UV selectively, thereby functioning as UV filter screens. Dust clouds could make climate control easier and cheaper by orders of magnitude than high-technology foil screens.

f. Cost-Benefit Assessment

It is necessary to compare the costs of a space screen program with (1) the benefits and (2) alternative measures. As noted above, the space screen costs could be readily offset if combined in a larger program with the satellite solar power and/or asteroid retrieval. Otherwise, the costs of the screen should be compared with the greenhouse damage that would be prevented. The total world value of inundated areas after greenhouse warming was estimated as 2.5×10^{12} in 1972 US dollars [5]. Counting desertification and wilderness losses, this can probably be increased to at least 10^{13}. Even without spinoffs, the solar screen would prevent these losses at 1% of the cost.

An alternative measure is to reduce CO_2 emissions. It is estimated that this could cost on the order of 10^{11} for several decades, i.e., on the order of 10^{12} - 10^{13} in total [1]. Again, the solar screen costs are smaller. Of course, prevention is more desirable, but reducing CO_2 emissions in an increasingly industrialized world would entail great sacrifices, including economic cutbacks in the industrialized nations and slowed development in the increasing over-populated developing nations.

These difficulties may be insurmountable, especially since they address problems in the future and the costs must be discounted accordingly. The rate of economic discounting was considered by Schneider as 7% per year, which computes to the effect that a damage of D dollars in 150 years is worth only D/26,000 in prevention costs at the present [5]. Quantitatively, the relation is expressed by equation:

$$C = d^{-t}D \qquad (1)$$

Here C is the cost of prevention, D is the future damage, d is the discounting rate and t is the time interval in years.

While discounting is valid qualitatively, this simple approach seems to underestimate the human concern for the future, and can lead to unreasonable conclusions. For example, extrapolating further, one may assume an ultimately stabilized world population of 10^{10} and $100,000 value per human life (comparable to current insurance practices). Then equation (1) suggests that the world community presently would not be willing to invest $1 to prevent the extinction of humanity in 510 years. This is demonstrably wrong since the US government alone is investing billions of dollars to isolate radioactive wastes for thousands of years, and resources on the order of 10^7 10^8 are spend annually to protect endangered species; both enterprises without any quantifiable economic returns.

Such considerations suggest that equation (1) is oversimplified and that values beyond discountable economic factors, or at least discountable on a different time-scale, should be included. A quantitative expression is given in equation:

310

$$C = P_pPD(d_v^{-t}V + d_D^{-t}D) \qquad (2)$$

Here P_p is the probability of the success of prevention; P_d is the probability of the damage; and V is a value beyond discountable economics. The equation assumes that even such values are or discounted at a very slow rate depending on the remoteness of the risk. (For example, how much would be spent presently to prevent the Sun from incinerating the Earth 5 billion years in the future?). Applying equation (2), we assume that the occurrence of a greenhouse effect is 50% probable, i.e., $P_d = 0.5$ and that the success of a space screen program is also 50% probably, i.e., $P_p = 0.5$. By a conservative estimate, a program that could prevent a devastating greenhouse effect any time within the next few centuries would be funded currently at the level of $10 billion, i.e., $V = \$10^{10}$, and $d_v = 1$, i.e., this value is independent of time on the scale of interest. The values of $D = \$10^{13}$ and $d_D = 1.07$ are used as above [5]. Equation (2) then suggests that preventing a greenhouse effect would justify an expenditure of 1.3×10^{12}, 6.5×10^{11}, 1.7×10^{11} and 5.4×10^{9} at 10, 20, 40 and 100 years in advance, respectively. Therefore, given a 50% probability of a greenhouse effect in 100 years, it would be justified at the present to spend $5 billion on a research program toward a space screen. An expenditure of the required $100 billion for a preventive space-based screen would be justified 40 year in advance. This time schedule is reasonable also for technical reasons, since deploying the screen would probably take several decades in any event.

Summary

The expected greenhouse effect could be prevented using a space-based solar screen, probably deployed about the Sun-Earth L1 point or in levitated Earth-synchronous solar orbit. The scale of this program would be comparable to the construction of a space colony for 10,000 people. Indeed, the solar screen program should be part of overall space development on that scale, also involving satellite solar power and asteroid retrieval, all with substantial benefits to Earth.

In the presence of space activities on that scale, deploying and maintaining a solar screen will be comparable to operating an aircraft carrier and associated naval fleet today: a substantial program but readily feasible. Even without such spinoffs however, cost-benefit considerations justify a substantial research program starting now, and deployment 40 years in advance of the onset of a severe greenhouse effect.

The space screen should not be considered as an alternative to environmental prevention. In medical analogy, the space screen is equivalent to symptomatic treatment. Preventive care or cure is preferable; but failing that, treating harmful symptoms is preferable to inaction. In particular, the space screen would be justified at any cost if positive feedback effects threaten with runaway heating that leads to an uninhabitable Earth.

The designed control of the climate could be comparable in impact to other major technologies, such as nuclear and biotechnology. Similar to those developments, it should be pursued prudently. The important point at this time is to realize that space-based control of the climate will become readily feasible within a few decades. This suggests the needs for further research and public debate.

Appendix A3.8

Journal of The British Interplanetary Society, Vol. 44, pp.135-138, 1991

A SPACE-BASED SOLAR SCREEN AGAINST CLIMATIC WARMING*

M. MAUTNER
Department of Chemistry, University of Canterbury, Christchurch 1, New Zealand.

*Paper submitted August 1988

Abstract

The expected global warming may be reversed by space-based screen that would intercept a fraction of the solar radiation incident on Earth. Warming by 2-5°C could be prevented by intercepting 3-7%, respectively, of the incident solar radiation. The screen may be constructed for example of a thin film, a grid of film supported by a mesh or fine-grained dust deployed in orbit about the Earth. For an average film thickness or particle radius of 0.001 cm, the required mass is 10^{14} - 10^{15} g, equivalent to a medium sized lunar mountain or asteroid. The material may b deployed and processed using existing technology, similar to methods proposed for space habitat construction. The cost may be substantially lower than the economic and human damage caused by a significant climate change.

Appendix A3.9

Paper on Lunar Gene Bank for Endangered Species

Journal of The British Interplanetary Society, Vol. 49, pp 319-320, 1996

SPACE-BASED GENETIC CRYOCONSERVATION OF ENDANGERED SPECIES

MICHAEL N. MAUTNER
Department of Chemistry, University of Canterbury, Christchurch 8001, and Department of Soil Science, Lincoln University, Lincoln, New Zealand

---

Genetic materials of endangered species must be maintained, for cryoconservation, permanently near liquid nitrogen temperatures of 77 K. Due to the instability of human institutions, permanent safety is best provided at storage sites that maintain passively the needed low temperatures, and provide barriers to access. The required conditions are available in permanently shaded polar lunar craters with equilibrium temperatures of 40 - 8- K, on the moons of Saturn, and unshielded storage satellites. A genetic depository can be incorporated readily into planned lunar programmes.

---

1. Introduction

Over the next 50 years, 15-30% of the estimated 5-10 million species may disappear due to human pressures. The losses include hundreds of vertebrate, hundreds of thousands of plant and over a million insect species [1]. The gene pools of many human ethnic groups are also threatened. For many animals, adequate conservation of habitats is unfeasible, and active breeding programs cover only 175 of the many thousand species threatened [2]. The genetic heritage of the living world, accumulated during aeons of evolution, is being wasted in a short period.

Against such losses, scientists are starting cryopreservation programs of genetic material in tissue samples, semen and embryos [3-6]. However, funding is already tenuous. During centuries, incidents of war, sabotage, disasters, economic depression or just a loss of interest are bound to happen. Even a brief disruption of the permanent refrigeration can destroy the samples, and this prospect discourages cryopreservation programs. Secure preservation requires remote sites immune to such disruptions, at locations where the samples can survive periodic abandonment and remain stable indefinitely at natural temperatures around 80 K.

2. Setting Forth the Suitable Sites

No terrestrial sites satisfy these requirements. However, the required equilibrium temperatures exist at lunar polar craters, or at more remote planetary sites that can be used for long-term backup storage. About 0.5% of the total lunar surface area is estimated to be permanently shadowed and remain below 100 K. Thermal balance calculations suggest steady-state temperatures of 40 - 60 K at the shadowed centers of polar craters [7]. The lunar sites are safe from earthquakes due to the low lunar seismic activity, with an average of five events per year with Richter magnitudes of only $2.2 < m < 4.8$ [7]. At these sites, burial in one meter of the regolith will protect against the solar wind, solar and galactic cosmic rays and micrometeorite impact [8].

Other planetary locations with equilibrium temperatures about 70-90K exist on the moons of Saturn [9]. The atmosphere of Titan with a surface pressure of 1.5 bars can protect from space vacuum, but prevents easy access and retrieval, and the possible surface cover of liquid N_2 is mechanically unfavorable for retrieval. Ease of retrieval favors small, low-gravity moons such as the high-albedo Enceladus with an average temperature of 70 K, or Phoebe, whose large distance from Saturn (1.3×10^5 km) and small size 220 km diameter) allows a small escape velocity.

Feasible deposits may be established also on storage satellites in Earth orbit. Shields for thermal insulation and from ionizing radiation can be provided, using shields equipped with passive attitude control by solar sail devices. More simple and secure, but harder to retrieve storage satellites, may be deposited in Solar orbit at about 10 au where equilibrium temperatures are around 80 K.

At lunar or planetary sites, samples will be subject to space vacuum, with at most 2×10^5 particles/cm^3 in the lunar environment [7]. Although some microbial material can survive space vacuum and low temperatures for several months, tissues from higher plants and animals should be encapsulated, in addition to the glycerol cryoprotector, against dehydration that can disrupt hydrophobic bonds in membranes, and induce conformational changes and crosslinks in DNA and proteins [10].

3. Requirements for Restoring Endangered Species

To restore an endangered species with viable genetic diversity, material from 20-30 unrelated individuals is regarded as minimum [11], and 0.1 g material from each founder individual can be sufficient using advanced genetic techniques. Using cryopreservation, it may not be necessary, or even desirable to extract DNA. Whole cells or small tissue samples will be easier to prepare for conservation, and to clone back later. To preserve microbial fauna, it is not even necessary (or possible) to identify each species now but microbial filtrates from various habitats such as soils, lakes, oceans and those living on other organisms, can be preserved.

The germplasm or tissue samples of one million endangered species can be accommodated in a payload of 2,000 kg. At expected future launch costs of $1,000 - 10,000/kg, the genetic heritage of one million endangered species can be stored in permanent safety at the cost of $2 to 20 million, or only a few dollars per species. Secure permanent space-based storage will require a smaller one-time cost than that required from more vulnerable terrestrial storage for only a few decades.

Lunar sites appear to best combine physical features with feasibility and security of access. The required transport of 2,000 kg is comparable to the proposal Lunar Cluster Telescope and less than the Large Lunar Telescope or the Lunar Synthesis Array [12]. Construction of a required shallow underground vault of 10 m^3 is less demanding technologically than the construction of lunar observatories or permanent bases. Therefore, the genetic storage program can be incorporated at little extra engineering or economic costs into lunar programs planned in the next 20-40 years. This time frame allows collection of the genetic material.

Even before the permanent storage sites are constructed, payloads of the genetic collections can be soft-landed into the cold craters, and deposited in shallow burrows by robots. When human return to the Moon, these collection can be transferred to the permanent depositories.

In addition to plant and animal species, the genetic material of human groups is also being lost as populations disperse and intermix. This material is important for the understanding of human evolution, history and biology. Genetic materials of vanishing human groups can be also preserved in space-based depositories.

4. The Philosophy of Space-Based Cryoconservation of Endangered Species

Conservation of endangered species is a popular program. Cryoconservation can add an important ethical component to the space programme and help raising public support. Conversely, the permanent safety of the genetic material can also make cryoconservation itself more attractive and fundable. Many nations may wish to participate to secure the genetic heritage of their unique biota and ethnic groups.

The genetic heritage unites and enriches the living world, and preserving it is a top ethical priority. Until habitat losses are controlled, cryoconservation may provide the best change to secure and eventually revive many endangered species. For this purpose, space-based depositories can provide the most cost-effective and secure means for permanent storage of irreplaceable genetic materials.

References

1. B A Wilcox, "1988 IUCN Red List of Threatened Animals", Gresham Press, Old Woking, Surrey, 1988, pp 12-16.
2. LM DeBoer, "Biotechnology and the Conservation of Genetic Diversity", eds H D M Moore, W V Holt and G M Mace, Claredon Press, Oxford, 1992, pp 5-16.
3. M E Soule, see ref. 2 pp. 225-234.
4. R A Adams, *Diversity*, **8**, 23 (1992).
5. J S Mattick, E M Ablett and D L Edmonson, "Conservation of Plant Genes", Academic Press, 1992, pp 15-35.
6. W V Holt, see ref 2, pp. 19-36.
7. B H Foing, Adv. *Space. Res.,* **14**, pp. 9-18, (1994).
8. T Entrenaz, J P Bibring and M Blanc, "The Solar System", Springer-Verlag, New York, 1990, p. 150.
9. See ref 8, p 226-229.
10. K Dose, *Adv. Space Res.,* **6**, pp. 307-312, (1986).
11. T J Foose, R Lande, R Flesness, G Rabb and B Read, *Zoo Biol.,* **5**, pp 139-146 (1986).
12. J P Wetzel and S W Johnson, *Adv. Space Res.,* **14**, pp. 253-257, (1994).
13. M Meot-Ner (Mautner) and G L Matloff, *JBIS*, **32**, p. 419, (1973).
14. M N Mautner, *JBIS*, **48**, p. 435, (1995).

Appendix A3.10

Infinitely Embedded Universes and Infinite Levels of Matter

In Chapter 6.5 we described a model of infinitely divisible matter and infinite epochs. This section describes the model somewhat more quantitatively.

The Main Features of the Model

1. The universe is expanding in an infinite series of epochs. Each epoch is progressively shorter as we look back, and the total duration of the infinite past epochs is a very short and finite sum of a convergent geometrical series. Each preceding epoch is increasingly hotter and denser, converging toward, but never reaching, a point in time and a state of infinite density and temperature.
2. Matter is divisible infinitely into smaller and smaller particles on lower and lower levels of magnitude, and infinitely compressible.
3. The infinite epochs of time and infinite levels of magnitude are related. Matter could be ever more compressed in past epochs if it is infinitely divisible and compressible.

These properties allow the universe to have a finite past duration and contents as observed, yet also to have an infinite past and to contain infinite particles, as may be expected. The model also provides a universal symmetry, in that Nature appears the same to any observer in any epoch of time and on any level of magnitude, as follows.

4. The universe will keep expanding through an infinite series of future epochs that will be ever longer, more dilute and colder.
5. In each epoch of expansion, every particle will separate beyond the event horizon of every other particle. Each particle will then become an expanding universe made of its sub-particles, until each particle becomes an expanding universe of its sub-particles, and so on in each epoch. By symmetry, the universe of every epoch was a particle interacting with other universe/particles in each preceding epoch.
6. Therefore this model provides universal Copernican symmetry. An observer on any level sees himself in the center of an infinite

series of smaller and of larger levels of magnitude. Each observer in any epoch sees himself in the center of ever shorter, denser and hotter past epochs and of ever longer, more dilute and colder future epochs. The observer sees himself in the center at any level and in any epoch because in an infinite sequence every point is the center.

Further, we shall note below that the model of infinite levels and epochs may also explain some aspects of relativity and quantum mechanics in conventional terms; indicate that Nature can be both deterministic and statistical, having infinite variables; and may allow self-propagating structured "life" through infinite epochs.

Indications that the Scales of Magnitude are Infinite in Both Directions

The shortest time that has physical meaning in our universe is 5.4×10^{-44} seconds, the Planck time during which light travels the shortest meaningful length, 1.6×10^{-35} m, the Planck length. At the other extreme, the longest meaningful time for material beings may be the time at which baryonic matter decays, with a suggested half-life of 10^{37} years (Adams and Laughlin, 1999). The time scale of biological events is on the order of seconds to years, about the center of the shortest and longest physically meaningful time-spans.

In dimensions of space, the smallest physically meaningful size is the Planck length, 1.6×10^{-35} m, and the largest meaningful size currently is the present size of the universe on the order of 1.4×10^{26} meters. In comparison, the biological units of length may be from the typical cell of 10^{-6} meters to the largest plants of 100 meters, centered about 0.01 meter. This is again near the center of the smallest and largest physically meaningful sizes.

In terms of mass, the smallest rest-mass of an ordinary particle is that of the electron, 9.1×10^{-31} kg (neutrinos may be much lighter). The largest meaningful mass with which we can interact directly is that of a galaxy, on the order of 10^{40} kg. In comparison, the mass range of direct human experience spans from milligrams (10^{-6} kg) to the mass of a large human, on the scale of 100 kg. This is again near the center of the smallest and largest physically meaningful masses.

It appears that the scales of our readily observed world, defined by the scales of biology, are near the center between the extremes of time, space and mass. General Copernican symmetry suggests that this is not a coincidence. We may appear to be in the center because we are looking toward horizons of scales that stretch from the infinitely small to the infinitely large. However, we can perceive only the scales of magnitude nearest to us with which we can exchange causal information.

Infinite Levels of Magnitude and Infinite Time Epochs

We may assume that all matter is infinitely divisible and compressible. It is indeed beyond imagination that all the stars, planets and galaxies were compressed into a volume much smaller than an atom on the point of a pin, an incomprehensibly small 1.6×10^{-35} meters. Nevertheless, this compression is consistent with physics and supported by observation. It would seem not less believable then that all matter could have been condensed by similar factors from epoch to epoch back in time, in an infinite sequence of epochs. In each earlier epoch time and space were much more compressed, events occurred on a faster time-scale (the pace of events may be scaled proportionally to the duration of the epoch). In each preceding epoch matter/energy was much denser and hotter, approaching but never reaching infinite compression, density and temperature. Each epoch lasted for shorter and shorter times, as seen from our scale (maybe very much shorter, say by a factor of 10^{80}). The total duration of this infinite sequence is a geometrical series that converges to a finite and very short time-span.

In the future, each epoch will last longer and longer and matter will be and colder and more dilute, approaching but never reaching a state of infinite cold and zero density. The duration of each future epoch is finite, but their sum is a divergent geometrical series that will always approach but never reach infinity.

In each epoch, an observer sees a finite universe made of finite matter/energy and a finite number of constituent particles. Each particle is made of sub-particles on a smaller level, maybe very many subparticles (say 10^{80}), each of which is made in turn of many sub-particles and so on. Each particle is a finite universe of sub-particles, which are divided again and again into more sub-particle/universes on infinite levels, so that ultimately the contents of each particle/universe approach an infinite number of infinitely lighter sub-particles. As the universe expands into future epochs, each particle will become an isolated universe made of its sub-particles.

Going back, each universe on any level was a compressed into a particle in a super-universe in the preceding epoch. There were an infinite number of preceding more compressed and massive universes, converging back in time toward an infinitely compressed and massive universe.

In each particle, the constituent sub-particles are separated by space and most of the matter is concentrated in small volumes. Proceeding to infinitely smaller levels, the density of energy/matter increases and approaches point-like particles of zero volume and infinite density. In this respect proceeding to smaller and smaller levels is similar to proceeding to earlier and earlier epochs where matter was increasingly hotter and more condensed.

Infinite Levels of Magnitude, Epochs, and Universes

We can state the preceding section more quantitatively. Assume that our universe u_0 exits in the time epoch E_0. The component particles of our universe are of level L_0 and they can interact with each other on this level. Each of these particles is made of a large but finite number n_0 of particles, say 10^{80} protons and electrons. Each particle may be made of n_{-1} (say 10^{100}) sub-particles of the next smaller level L_{-1} and therefore our universe contains $n_0 \times n_{-1} = 10^{180}$ sub-particles of level -1. In turn, each of these sub-particles is made of n_{-2} (say 10^{60}) sub-particles on the next lower level L_{-2} and so on. In this model our universe is made $n_0 \times n_{-1} \times n_{-2} = 10^{240}$ sub-particles of level L_2. If there are infinite levels, our universe contains a number of particles given by the product πn_i (i = 0 to infinity) = $n_0 \times n_{-1} \times n_{-2} \times n_{-3} \times n_{-4}$ .... to infinity, i.e., an infinite number of sub-particles. In other words, our observable universe is finite, containing a finite number n_0 of components on level L_0 that can communicate with each other. However, our universe is also infinite, as it contains infinite components on infinite sub-levels.

As the universe expands, each particle will recede beyond the event horizon of all other particles and become an expanding universe u_1 of epoch E_1 made of its sub-particles of level L_{-1}, which in turn will become the expanding universes u_2 of epoch E_2 made of their sub-particles of level L_{-2} and so on to infinity.

Similarly, our universe u_0 of level L_0 would have been then a particle of level L_1 in a universe u_{-1} of epoch E_{-1}. This universe contained a large number n_1 (say 10^{80}) of such universe/particles which could have interacted and exchanged information on that level L_1. Since this universe u_1 contained our universe, it already contained infinite sub-particles on infinite sub-levels, but it contained n_1 times more particles. This "super-universe" was in turn a particle in a universe u_{-2} of epoch E_{-2} which contained the next larger L_2 which contained n_2 such super-universes and so on to infinitely more remote past universes of past epochs. At the end of each epoch of expansion, the component universe/particles of universe u_{-1} separated beyond communication and became separate expanding universes, as our particles will be at the end of the present epoch.

In general, in epoch E_i the universe u_i contains n_i particles of level L_i. All of these particles were combined as a particle of level L_{i+1} in the preceding epoch E_{i-1} as a particle in universe u_{i-1} that contained n_{i-1} such particles. Similarly, each particle of level L_i of epoch E_i will separate beyond their mutual event horizons in the following epoch E_{i+1} and become a universe u_{i+1} made of n_{i+1} particles of level L_{i-1}. Our universe was a particle of the preceding epoch, and each of our particles will become a separate universe in the next epoch, and so on to infinity.

In each epoch there may be moderate conditions when the density of energy and the "temperature" allow particles to form structures, including self-propagating "life". The physical constants of each epoch may be determined by its only initial property, its mass/energy. Life in one epoch may then configure the initial mass/energy of the next epoch so that its physical constants will allow life. Our finite universe may be part of an infinite chain of existence, and our family of life may be part of an infinite chain of life.

Infinite Past Epochs in Finite Time

The preceding sections looked at the compression of matter in a particle in infinite steps during finite time. We may see such an infinite sequence of compressions if we look back at the history of the universe. This infinite sequence of past universes could have happened in a finite, indeed very short time.

On each smaller level, events are faster and time is more compressed. For example, the orbital period of two particles about each other on level L_{-1} may be orders of magnitude shorter than the classical orbital period of an electron about a proton on our level.

Measured on our time-scale, the duration of the inflation was on the order of 10^{-35} seconds. This was the duration of the previous epoch E_{-1} of the universe. Similarly, observers in that epoch would have seen an inflation that started their own Big Bang, which would have been shorter again say by a similar factor, lasting 10^{-70} seconds as observed from our scale. This was the duration of the E_{-2} epoch. In turn, observers in that epoch would have seen the duration of epoch E_{-3} that preceded them similarly shorter as 10^{-105} seconds on our scale. The total duration of time that preceded us is therefore $10^{-35} + 10^{-70} + 10^{-105} + 10^{-140}$ seconds and so on, an infinite geometrical series converging to but never reaching time zero. Although there was an infinite sequence of past epochs of expanding universes, their total time span is short and finite, given by the sum of a convergent geometrical series, $(10^{-35}/(1-10^{-35}))$ seconds, very close to 10^{-35} seconds of the preceding epoch alone.

Assume that the present epoch E_0 of universe u_0 from the Planck time to the decay of matter will exist for t_0 units of time, say 10^{37} years. Assume that the length of each preceding epoch becomes shorter by a constant factor $f < 1$ (of course, the factor may in fact vary). Then epoch E_{-1} lasted for ft_0, epoch E_{-2} lasted for f^2t_0 seconds, and epoch E_{-i} lasted for t_0f^i seconds. The total duration of all the infinite past epochs is then given by the sum of a geometrical series as $t = t_0[1/(1-f)]$. If the scaling factor between epochs is very large, for example if $f = 10^{-35}$ as above, then t is practically equal to t_0, i.e., the total length of the preceding epochs is negligible compared with our epoch. Even though there were an infinite number of preceding epochs, their duration converges to a finite length of time, and the whole past history of infinite nested Big Bangs expanding into universes is much shorter than the history of our own present universe as we see it from our level. An observer looking back from any previous epoch would have the past similarly shorter than his own epoch.

In the other direction, the next epoch E_1 may be similarly longer than our epoch lasting t_0/f seconds and any subsequent epoch E_i will be progressively longer lasting t_0/f^i seconds. The dilute, cold, slow paced and long-lasting expansion of one epoch will be seen as the dense, hot, fast-paced and brief Big Bang to observers in the next epoch. For an observer in a future epoch, the 10^{37} year expansion of our universe will have been one of the immensely brief hot epochs that preceded their Big Bang.

Interestingly, the first instant in our universe was a period of accelerating inflation. It appears that our universe is now again expanding at an accelerating rate. Maybe our expansion is really the inflation of the next epoch.

As our universe expands, each of our particles will recede beyond the event horizon of all the other particles and become an expanding universe. The decay of baryonic matter may be the dissolution of our particles into their sub-particles in this process.

Infinitely Compressible Volume

The volume occupied by ordinary matter is a small fraction of the volume of the universe, since most of the volume of galaxies, solar systems and atoms is occupied by vacuum. Assume that the particles of any level L_i become compressed and the vacuum amongst them is eliminated. The volume occupied by particles decreases by a factor $f_i < 1$ after the compression. If the compression is carried out at each level from L_o down to level L_{-n} and if the compression factor in each step i is $f_i < 1$ then the remaining volume occupied by matter after n steps is $V_n = V_o \Pi f_i$ (i = 0 to n). After infinite steps of compression to an infinitely small level, V_n as given by this product approaches zero as i approaches infinity. Therefore if matter is infinitely divisible into particles that are separated by vacuum on each level, then it is infinitely compressible, the volume occupied by matter is zero and matter is in fact made entirely of vacuum.

If matter were compressed in this manner, eliminating the vacuum amongst the particles at each level without the loss of matter/energy, then after infinite compressions on infinite levels matter would assume infinite density and temperature. In practice, matter would become a black hole at some step. However, each step of compression would in fact involve the loss of radiated energy and all the initial apparent rest-mass would turn into energy, as described below.

Density and Temperature

In our model, matter is concentrated in smaller and smaller particles in going to increasingly smaller levels, and the density of matter in the particles increases. For example, the density of matter in the total volume of the Solar System is much smaller than the density of same mass concentrated in the volume of the Sun and the planets. Similarly, we can calculate the volume of a ground-state hydrogen atom as 4.2×10^{-30} m^3 and its density as 4.0×10^2 kg m^{-3}, but most of the mass is concentrated in the proton that has a much smaller volume of 4.2×10^{-45} m^3 and greater density of 4.0×10^7 kg m^{-3}, respectively. The density of a matter in the particles of level L_i becomes successively greater as we resolve matter to smaller and smaller units concentrated in smaller and smaller volumes, eliminating the vacuum that separates particles on each level.

Matter can be assigned an equivalent radiation temperature, which is the temperature at which black-body radiation concentrated in the same volume would have the same mass/energy. If the density of matter concentrated in particles increases on each smaller level, then the equivalent temperature of matter in the particles increases according to the relation between energy density and temperature as given by $\rho_{radiation} = \alpha T^4 c^{-2}$, where $\rho_{radiation}$ is the radiation density, α is the radiation constant, T the absolute temperature and c the speed of light. For example, we can calculate the equivalent temperature of a hydrogen atom and the proton from their densities of 4.0×10^2 and 4.0×10^7 kg m^{-3} as 4.7×10^8 K and 8.3×10^9 K respectively.

Going to increasingly smaller levels, the densities of particles and their equivalent black-body temperatures increase. If the scale of length decreases by a factor f_i from level to level, the volumes decrease and the densities increase as f_i^3 and the equivalent temperature of particles increases as $f_i^{3/4}$ from level to level.

Going in the other direction, matter becomes more dilute and cold as we go to larger scales. For example, the present density of baryonic matter in the universe is 8.8×10^{-29} kg/m^3 and its equivalent black-body temperature is 10.2 K, close to the actual background radiation temperature of 2.7 K. Going to each higher level matter becomes more dilute and its equivalent temperature approaches but never reaches absolute zero.

Whether density and temperature are seen as very low or very high depends on the observer's level. Even an inconceivably high but finite temperature of 10^{100} K will seem as immensely cold from a smaller and hotter level and even a temperature only minutely higher than absolute zero such as 10^{-100} K will seem as enormously hot from yet larger and colder level in an infinite sequence of levels, epochs, densities and temperatures.

Mass and Energy

The apparent rest-mass of a system, as observed for example by its gravitational effect on a remote object, is equal to the rest-mass as observed internally in the system, plus the mass-equivalent of its energy.

For example, the mass of a solar system as observed from another star is equal to the rest-masses of the Sun and of planets, asteroids and comets that orbit it, plus the mass equivalents of their orbital and thermal energies. In turn, the rest-masses of atoms that make up these objects are equal to the rest-masses of the nuclei and electrons plus the mass equivalent of their energy. Similarly, the rest mass m_i of an object as observed from level L_i is the sum of the rest masses of its components plus the mass equivalent of their energies as observed from the next smaller level L_{i-1}.

An object with rest mass m_i as observed on level L_i is observed on the next smaller level as a collection of particles with a total rest-mass of $f_i m_i$ and with the rest of the mass equal to $(1 - f_i) m_i c^2$ observed as energy. Assume for simplicity that the factor $f_i < 1$ is equal in going between every adjacent level. Starting with a particle with a mass m_0 at our level L_0 the observed remaining rest-mass at a smaller level L_n is $m_0 \Pi f_i$ ($i = 0$ to n-1) plus the remaining original mass appearing as $e = m_0 c^2 \Pi(1 - f_i)$ ($i = 0$ to n-1)c^2 of energy. If we proceed to infinitely smaller levels, the remaining inertial mass as observed from each smaller level decreases to zero. If on all levels matter is made of moving particles, then all the mass observed on any level L_i as rest-mass will appear as kinetic energy when observed from an infinitely small level.

In other words, if matter is infinitely divisible and the constituent particles on each level contain energy, then a smaller fraction of the original mass m_0 is observed as rest-mass, and more is observed as energy from increasingly smaller levels. All of the mass observed on any finite level is observed as energy from an infinitely small level.

Compression on Infinite Levels, and the Conversion of Mass into Energy

What happens if the system is compressed or collapses stepwise on each level to the maximum density allowed by the particles of that level, and the energy content on each level is released as radiation? The remaining rest-mass is $m_0 \Pi f_i$ ($i = 0$ to infinity) but if all $f_i < 1$ then the product approaches zero as i approaches infinity. The remaining rest-mass vanishes, and all of the mass m_i observed on level L_i will have been converted into $m_i c^2$ energy.

The compression of the particles on successive smaller levels can be increasingly faster. The total time for infinite steps of compression on infinite levels is the sum of a series that converges to a finite length of time.

For example, if compression on level L_0 requires t_0 units of time and the scaling factor of time is $f_i < 1$ from level to level, then compression on the next sub-level requires $f_0 t_0$ units of time. Starting from level L_0, the duration of compression on level L_{-n} is $t_0 (\Pi f_i)$ ($i = 0$ to n) units of time. The total time required for the n steps of compression is $t_0 \Sigma (\Pi f_i)$ ($i = 0$ to n) units of time. For example, if the factor of time scales f is constant from step to step, the time required for the process is the sum of a series $t_0(1 + f + f^2 ....) = t_0/(1 - f)$ time units. The required time for the an infinite series of compressions is the sum of a geometrical series, which is finite. If $f \ll 1$, then $(1 - f)$ is about unity, and the entire process occurs nearly in the same time t_0 as the first step. In other words, an infinite number of steps of compression with the release of all of the rest-mass of level L_i can be completed in finite time. This process can provide for the conversion of mass into energy in a short finite time as seen from our level of magnitude.

This mechanism assumes that on each level matter contains kinetic energy as observed from that level, such as the kinetic energy of particles revolving about each other on every level. When matter is compressed, this energy is released as radiation. If the distribution of kinetic and potential energy obeys the virial theorem on every level, then this process may lead to the $e = mc^2$ relation.

The successive steps of compression would take shorter and shorter times, and would leave increasingly more dense and hotter matter. In these respects the infinite stages of collapse of matter is similar to going back to increasingly shorter, denser and hotter past epochs of the universe.

The Scaling Factors

The model of infinite levels requires that the scales of objects vary from level to level. For example, the sizes of sub-particles of our particles are so small that we cannot interact with them directly. We cannot affect, or be affected by a particle that is 10^{80} times smaller and lighter than a proton; neither can such a particle alone affect events on our scale. Further, along with the scaling factor of size there can be also a scaling factor of time and of the pace of events. For example, the orbital periods of particles 10^{80} times lighter than the proton may be shorter by a similar factor than the classical orbital period of an electron about a proton. Similarly, the pace of events in past epochs may have been faster by similarly large factors than in our epoch. It may be proportional to the total duration of that epoch, or it may be a function of the density and temperature of matter on each level of matter and in each epoch.

As an example of scaling factors, the ratio between the Planck time (10^{-35} seconds) and the lifetime of matter as defined by the lifetime of protons (3×10^{44} seconds) is about 10^{80} and the ratio between the mass of the proton (1.67×10^{-27} kg) to the mass of the universe (10^{52} kg) has a similar ratio. If these examples are representative, each epoch may span 80 orders of magnitude in space, time and scales of magnitude. The scaling factor from one level of magnitude to another may be very large.

The scales of space, time, mass, density and temperature may or may not vary together by the same factor by equal factors as we proceed from one level of matter to another or epoch to epochs. Moreover, these scaling factors may or may not be the same at each step in the infinite sequence of levels or epochs. It was assumed for simplicity in some of the preceding sections that these factors are constant, but the results apply qualitatively regardless of the scaling factors.

Quantum Effects

Quantum events behave statistically and statistical behavior usually reflects large numbers. The fact that quantum events are statistical may therefore reflect the statistics of large numbers of sub-components of each particle. For example, when a particle disintegrates, this may occur because enough energy or sub-particles are concentrated statistically in one volume to cause dissociation. This is the well established mechanism of the dissociation of molecules.

All particles are spread out in space, and in quantum mechanics any particle can be found anywhere in the universe with a finite probability. This would be possible if the sub-particles that compose each particle form a cloud that can spread out throughout the universe.

Maybe when the edges of the sub-particle clouds meet and inter-penetrate, this creates an area of high density. The rest of the sub-particle clouds condense into this dense center similarly to the collapse of interstellar clouds to form stars. This may provide a mechanism for the collapse of wave-functions in quantum mechanics. The sub-particles of the interacting particles can mix and divide to form new particles.

It would be possible for the sub-components of each particle to flow and explore all possible paths and chose the lowest energy path as proposed by Feynman.

The time-scales of events may be much shorter and events much faster on smaller sub-levels of matter, as discussed above. Sub-particles on smaller levels may communicate with each other much faster than on our level. In other words, transfer of energy on smaller levels may be much faster than on our level and the speed of light on smaller levels may be much greater than on our level.

This would also allow the collapse of wave-functions when particles interact to form new particles to be instantaneous as observed in quantum mechanics. Indeed, we observe such quantum events, and the transition between quantum states as instantaneous.

The compressed time-scales on smaller levels would also allow sub-particles to communicate at a speed that appears to us as instantaneous even when they are separated by large distances. For example, assume the on the next smaller level the speed of light is 10^{80} c, faster than the speed of light on our scale by eighty orders of magnitude. Then a particle of that sub-level can communicate with another particle 14 billion light-years or

4×10^{17} light-seconds away in 4.4×10^{-63} seconds, much faster than the scale on which we can resolve time. Such faster-than-light communication among particles is consistent with the evidence for quantum entanglement.

In summary, matter composed of many sub-components would explain the statistical nature of quantum events. If the time-scales of events are also compressed on smaller levels of matter, we could also explain instantaneous transitions between quantum states, the instantaneous collapse of wave-functions, and instantaneous communication between entangled particles. They are classical events on time-scales that are finite but too short to observe on our level. These sublevel events may explain some quantum processes in classical terms, and avoid infinitely fast quantum events by events that are very short but finite in time.

Causality

Although we cannot see events on sub-levels of matter, events on those scales may affect us. Matter made of many particles on many levels can exhibit *statistical determinism*: the future of an assembly of particles may be predictable, although not the future of each particle. In any epoch and on any level, the future may be predictable in its general outcome.

An observer who could measure the states of all of the sub-components of matter on all levels could apply the laws of physics to calculate the future. However, without such knowledge, the behavior of particles, as determined by their sub-particles, will appear to be statistical. A future that is determined causally by an infinite number of variables is on the borderline between deterministic and causal.

Can there be a causal interaction between consecutive epochs of the universe? We cannot interact with the previous epoch when the universe was smaller than the Planck scale. The laws of physics may be constant on every level and in every epoch. However, the physical constants of each epoch may be determined by the only physical property at the beginning of an epoch of expansion, which is the total mass/energy of the universe of that epoch. Maybe it is possible then that life in one given epoch can configure matter so that the initial mass/energy of the next epoch will lead to physical constants that also allow structured life.

Structure and Life

In each universe of each epoch there may be a temperate period when the temperature or energy density allows the forces and particles of that epoch to form structured matter. These conditions can allow the formation of self-propagating "life" in each epoch, and an infinite sequence of Life may continue in an infinite sequence of epochs.

In a classical paper, Freeman Dyson postulated in 1979 a scaling rule for life. As the universe gets colder and more dilute, the pace of events can slow down. A cognitive entity can be supplied with enough energy and can keep "thinking" and creating entropy forever, if the pace of its thought events slows down with the decreasing temperature.

This principle may be generalized in the other direction, to "life" in the hot, condensed and brief Big Bang and in even hotter, denser and briefer epochs that may have preceded it. As for the future, Dyson asked if an ever slower paced organized, cognitive "life" could last forever. We may ask if ever more fastly paced organized, cognitive "life" could have existed forever in past universes.

We may apply a scaling rule to earlier epochs. At the earliest time of our epoch, the Planck time of the universe, the mass of the universe, 10^{52} kg, was condensed in the Planck volume of $(4\pi/3)\lambda^3 = (4/3\pi) \times (1.6161 \times 10^{-35})^3$ m^3 = 1.76×10^{-104} m^3, giving a density of 5.7×10^{156} kg/m^3 and an equivalent black-body temperature, from the Stefan-Boltmann law, of 9.1×0^{46} K. Particles on our level could have flickered into and out of existence in very short time, say 10^{-40} seconds. However, this would still be a very long time if the scale of time was reduced, say, by a factor of 10^{80}. A "thought event" of ten milliseconds in the present epoch would have been shortened to 3×10^{-47} seconds in the preceding hot and fast-paced epoch.

Through similar scaling factors, there could be ever more fast-paced life in every preceding epoch and ever more slow-paced life in every cold future epoch. Similarly, there could be "life" in every sub-world of every particle on any level of magnitude.

Certainly thinking entities in preceding and future epochs were not made of current matter. However, different forces operate on each level of magnitude. If the forces on a particular level are strong enough, they can overcome disruption. There may be the organized life in infinite levels of magnitude.

The constants of physics in every epoch may be determined by the mass/energy of the universe of that epoch. In this case, life in one epoch may configure matter in the next epoch to also allow structured life.

Given infinite levels of magnitude, we may be part of an infinite expanse of life on infinite levels. If one epoch can configure matter in the next epoch, we may also be part of an infinite chain of self-propagating life in infinite epochs.

Conclusions

Ideas about worlds within worlds, infinitely divisible matter and a universe of infinite duration date back to antiquity. Our model shows that these notions are related to each other, that they can be combined into a framework consistent with science, and even provide mechanisms for processes in relativity and quantum mechanics.

Assuming that matter is infinitely divisible and compressible allows an infinite sequence of past universes in a finite brief time that preceded our universe. It also allows the unfolding of infinite future universes. The model satisfies a universal Copernican symmetry. Every observer at any level of magnitude is looking into an infinite sequence of smaller and smaller and of larger and larger levels. Any observer in any epoch is looking back into an infinite sequence of ever more compressed, hotter and short-lived epochs and looking forward to an infinite sequence of ever more dilute, colder and longer epochs.

The model may provide a mechanism for the relativistic conversion of mass into energy. It may also explain why quantum particles behave statistically, allows quantum particles to spread over the universe and be entangled through instantaneous communication. It implies a universe controlled by an infinite number of variables on infinite levels that is borderline between deterministic and statistical physics.

The model postulates levels of magnitude and epochs of time with which we cannot exchange information. It or may have consequences that can be tested experimentally. It allows the universe to be finite as is observed yet also infinite as we expect nature to be.

It is comforting to think of an infinite sequence of universes, each of which may contain its own form of "life". It is comforting to think of an infinite chain of worlds within worlds that may harbor structured matter and "life". It is comforting to be part of an infinite chain of Life.

Nature may hide infinity in a sequence of sub-levels that we cannot communicate with but that can affect us. It hints at an infinite sequence of universes within particles and universes that are, or were, particles in super-universes. Nature hints at an infinite series of ever shorter, fast-paced, more dense and hot past epochs that converge to a finite point of origin in time, and to an infinite sequence of ever longer, slow-paced, more dilute and colder future epochs that expand to infinity. Nature may then allow infinite life in worlds-within-worlds on infinite levels of magnitude, and infinite chain of self-propagating life within an infinite succession of epochs.

*F*or us in the present, it is comforting to think that our finite universe is part of an infinite chain of existence of ever-expanding universes, and that our finite life is part of an infinite chain of "life" that preceded us in previous epochs and that will follow us in future epochs. Our descendants through the epochs may seek to configure Nature so that Life can continue forever in infinite future epochs. Although causal measurements by finite beings cannot confirm them, observable science hints of such an infinite chain of existence and life.

As for us in this universe, we must be thankful for the knowledge that Nature does allow us. We learned that we are part of the family of organic life that is unique and precious. Nature will offer us to explore much more and to construct a great diversity of living structures. If we expand and propagate life, we shall have an immense future. Our descendants will achieve deeper knowledge and may seek eternity. In our descendants, our human existence will fulfil a cosmic purpose.

BIBLIOGRAPHY

There exists a large literature on future life in space. The following selected bibliography is particularly pertinent. Many references can be found also on the internet.

1. Adams, Fred and Laughlin, Greg, *The Five Ages of the Universe* (New York: Touchstone, 1999).
2. Anders E., Prebiotic Organic Matter from Comets and Asteroids, Nature 1979, 342, 255-257.
3. Arrhenius, S., 1908. Worlds in Making, Harpers, London
4. Baldauf S. L., Palmer, J. D. and Doolittle, W. F., "The Root of the Universal Tree and the Origin of Eukaryotes Based on Elongation Factor Phylogeny." *Proceedings of the National Academy of Sciences of the United States of America* 93 (1996): 7749-7754
5. Bowen, H. J. M., 1966. Trace Elements in Biochemistry, Academic Press, New York.
6. Brearley, A. J., and Jones, R. H., 1997. Carbonaceous Meteorites, in Planetary Materials, J. J. Papike, ed., Reviews in Mineralogy, Mineralogical Society of America, Washington, D. C. Vol. 36, pp. 3-1 - 3-398.
7. Chyba, C. F. and McDonald, G. D., "The Origin of Life in the Solar System: Current Issues", Annu. Rev. Earth Planet. Sci., 1995, 23, 215-249.
8. Clark Arthur C., *Profiles of the Future* (New York: Warner Books, 1984), pp 210.
9. Crick F. H. and Orgel, L. E., Directed Panspermia, Icarus 19, 341, 1973.
10. Crick Francis H., and Leslie E. Orgel, "Directed Panspermia", *Icarus* 19 (1973), 341-348
11. Criswell, David R., "Solar System Industrialization: Implications for Interstellar Migration", in *Interstellar Migration and Human Experience,* Ben R. Finney and Eric M. Jones, editors (Berkeley: University of California Press, 1985), p. 50-87.
    Davis, Paul C. W., *The Mind of God: The Scientific Basis for a Rational World* (New York: Touchstone Books, 1992), pp. 194
12. Des Jardins Joseph R., *Environmental Ethics: An Introduction to Environmental Philosophy* (Belmont: Wadsworth, 1997)
13. Dyson Freeman, *Infinite in All Directions* (New York: Harper and Row, 1988).
14. Dyson, F., 1979b. Disturbing the Universe, Harper and Row, New York.
15. Dyson, Freeman, "Time without end: Physics and biology in an open universe," *Rev. Modern Phys.* 51 (1979): 447-468.
16. Finney, Ben R., "Voyagers Into Ocean Space" in *Interstellar Migration and Human Experience,* Ben R. Finney and Eric M. Jones, editors (Berkeley: University of California Press, 1985), p. 164-180.
17. Fogg Martin J., "Terraforming: A Review for Environmentalists," *The Environmentalist* 13 (1993) 7-12.
18. Gibbons Ann, "Which of Our Genes Make Us Humans", *Science* 281 (1998): 1432-1434
19. Gribbin, John and Rees, Martin, *Cosmic Coincidences* (New York and London: Bantam Books, 1989), pp. 269

Hargove, E. C. (Ed.), 1986. Beyond Spaceship Earth: Environmental Ethics and the Solar System, Sierra Club Books, San Francisco.

20. Haynes, R. H., and McKay, C. P., 1993. The Implantation of Life on Mars: Feasiblity and Motivation, Adv. Space. Res. 12, 133-140.

21. Hart Michael H., "Interstellar Migration, the Biological Revolution, and the Future of the Galaxy," in *Interstellar Migration and Human Experience,* Ben R. Finney and Eric M. Jones, editors (Berkeley: University of California Press, 1985), p. 278.

22. Hartmann, Wiliam K., "The Resource Base in Our Solar System", in *Interstellar Migration and Human Experience,* Ben R. Finney and Eric M. Jones, editors (Berkeley: University of California Press, 1985), p. 26-41.

23. Hoyle F, and Wickramasinghe, C., Lifecloud: the Origin of Life in the Universe, J. M. Dent and Sons, London, 1978.

24. Lewis, J. S., 1996. Mining the Sky, Helix Books, Reading, Massachusetts.

25. Lewis, J. S., 1997. Physics and Chemistry of the Solar System., Academic Press, New York.

26. Mauldin J. H., "Prospects for Interstellar Travel", AAS Publications, Univelt, San Diego, 1992.

27. Mautner M and Matloff, G. L., A Technical and Ethical Evaluation of Seeding the Universe. Bulletin American Astronomical Soc. 1977, 9, 501.

28. McKay, C. P., 1992. Does Mars Have Rights? An Approach to the Environmental Ethics of Planetary Engineering, in D. MacNiven, (Ed.), Moral Expertise, Routledge, New York, pp. 184-197.

29. O'Leary, B. T., 1977. Mining the Apollo and Amor Asteroids, Science 197, pp. 363.

30. O'Neill Gerard K., *The High Frontier.* (New York: William Morrow, 1977).

31. O'Neill, G. K., 1974. The Colonization of Space, Physics Today, 27, pp. 32-38.

32. Rees, M. Our Final Hour: A Scientist's Warning: How Terror, Error, and Environmental Disaster Threaten Humankind's Future In This Century On Earth and Beyond. Basic Books, 2003.

33. Rynin N. A., *K. E. Tsiolkovskii: Life, Writings, and Rockets.* (Vol. 3, No. 7 of Interplanetary Flight and Communication. Leningrad Academy of Sciences of the U.S.S.R. Translated by the Israel Program for Scientific Translations, Jerusalem, 1971). "The Earth is the cradle of the human mind, but one cannot live in the cradle forever".

34. Schweitzer Albert, *Out of My Life and Thought* (New York: Holt, 1990) pp. 131. "(being good:) to preserve life, to promote life, to raise to its highest value life which is capable of development"

35. Shklovskii and Sagan, C., "Intelligent Life in the Universe", Dell, 1966.

36. Starchild, Adam, *Science Fiction of Konstantin Tsiolkovsky* (New York: International Specialized Book Services; 2000)

37. Taylor Paul W., *Respect for Nature: A Theory of Environmental Ethics* (Princeton: Princeton University Press, 1986), p. 45.

38. Valentine, James W., "The Origins of Evolutionary Novelty and Galactic Colonization", Ibid., p. 266-276.

Relevant publications by the present author
Directed Panspermia

1. Mautner, M. and Matloff, G., "Directed panspermia: A technical evaluation of seeding the universe" Ninth Annual Conference of the American Astronomical Society; Bull. Astr. Soc., 1977, 9, 501.
2. Mautner, M. and G. L., Matloff, "Directed Panspermia: A Technical and Ethical Evaluation of Seeding Nearby Solar Systems. J. British Interplanet. Soc. 1979, 32, 419-423.
3. Mautner, M. N.," Directed panspermia. 2. Technological advances toward seeding other solar systems, and the foundations of panbiotic ethics", J. British Interplanetary Soc., 1995, 48, 435.
4. Mautner, M. N., "Directed panspermia. 3. Strategies and motivation for seeding star-forming clouds", J. British Interplanetary Soc. 1997, 50, 93.
5. Mautner, M. N., "Society for Life in Space (SOLIS) (Interstellar Panspermia Society)", 1996. In "Marsbugs", J. Hiscox and D. Thomas, eds, 1995. E-mail: solis@eco88.com

Astrochemistry

1. Mautner, M., "The temperature dependencies of some ion-molecule reactions of astrochemical interest." Origins of Life 1975, 6, 377.
2. Mautner, M., "Ion-molecule condensation reactions: A mechanism for chemical synthesis in ionized reducing planetary atmospheres." Origins of Life 1978, 9, 115.
3. Milligan, D.; Wilson, P. F.; Freeman, C.; Mautner, M. N.; McEwan, M. J.,"Reactions of H_3^+ with hydrocarbons and nitrogen compounds, and some astrochemical implications". J. Phys. Chem. A 2002.

Space Technology for Climate Control and Conservation

1. Mautner, M., "Climatic Warming", *Chem. Eng. News,* June 1989 p. 33.
2. Mautner, M., "Deep-space solar screens against climatic warming: technical and research requirements." in "Space Utilization and Applications in the Pacific", Proceedings of the Third Pacific Basin International Symposium on Advances in Space Science, Technology and Applications, American Astronautical Society, November 6-8, 1989, Los Angeles. P. M. Bainum, G. L. May, T.

Yamanaka and Y. Jiachi, ed. p. 711. AAS paper 89-668.
3. Mautner, M.; Parks, K., "Space-based control of the climate." Proceedings of "Space 90: Engineering, Construction and Operations in Space" Proceedings of Conference of the ASCE/AIAA, April 23-26, 1990, Albuquerque, New Mexico. J. W. Johnson and J. P. Wetzel, ed. vol 2, 1159.
4. Mautner, M., "A space-based screen against the greenhouse effect." J. Brit. Interplanetary Soc., 1991, 44, 135.
5. Mautner, M. N., "Space-based genetic depositories for the cryopreservation of endangered species", J. British Interplanetary Soc., 1996, 49, 319.

Planetary Resources and Astroecology

1. Mautner, M. N.; Leonard, R. L.; Deamer, D. W., "Meteorite organics in planetary environments: Hydrothermal release, surface activity, and microbial utilization". Planetary and Space Science, 1995, 43, 139.
2. Mautner, M. N., "Biological potential of extraterrestrial materials. 1. Nutrients in carbonaceous meteorites, and effects on biological growth." Planetary and Space Science, 1997, 45, 653-664.
3. Mautner, M. N.; Conner, A. J.; Killham, K.; Deamer, D. W., "Biological potential of extraterrestrial materials. 2. Microbial and plant responses to nutrients in the Murchison carbonaceous meteorite." Icarus 1997, 129, 245-257.
4. Mautner, M. N., "Formation, chemistry and fertility of extraterrestrial soils: Cohesion, water adsorption and surface area of carbonaceous chondrites. Prebiotic and space resource applications." Icarus 1999, 137, 178 – 195.
5. Mautner, M. N. and Sinaj, S., "Water-extractable and exchangeable phosphate in Martian and carbonaceous chondrite meteorites and in planetary soil analogues." Geochim. Cosmochim Acta 2002, 66, 3161-3174.
6. Mautner, M. N.,"Planetary bioresources and astroecology. 1. Planetary microcosm bioassays of Martian and meteorite materials: soluble electrolytes, nutrients, and algal and plant responses." Icarus 2002, 158, 72-86.
7. Mautner, M. N., "Planetary resources and astroecology. Planetary microcosm models of asteroid and meteorite interiors: electrolyte solutions and microbial growth. Implications for space populations and panspermia."Astrobiology 2002, 2, 59-76.

Biographical Note The author served on the faculty of Rockefeller University, as a research scientist at the National Institute of Standards and Technology, and is currently Research Professor of Chemistry at Virginia Commonwealth University and Senior Fellow at the University of Canterbury and Lincoln University. He published numerous articles and book chapters in physical chemistry, astroecology, biophysics and space science. He taught adult education courses on science, society and the future, the author of popular science articles in "The Futurist" and "Spaceflight", and is a member of the Editorial Board of "Astrobiology". He has researched and advocated directed panspermia to seed the galaxy since 1977, and in 1996 founded the internet-based Society for Life in Space (SOLIS) - The Interstellar Panspermia Society.

Seeding the Universe with Life

www.ingramcontent.com/pod-product-compliance
Lightning Source LLC
Chambersburg PA
CBHW022052210326
41519CB00054B/322